AF520623

Participatory Rural Appraisal :

A Case Study, Approach and Implementation

Participatory Rural Appraisal:

A Case Study, Approach and Implementation

Project Leaders

Abhijit Khadatkar
ICAR-Central Institute of Agricultural Engineering, Bhopal (MP), India

Rameshwar Prasad Sah
ICAR-National Rice Research Institute, Cuttack (Odisha), India

Rajib Deb
ICAR-Central Institute for Research on Cattle, Meerut (UP), India

Sandeep Kumar, G.M.
ICAR-Indian Institute of Horticultural Research, Bangaluru (Karnataka), India

Shantanu Basak
ICAR-Central Institute for Research on Cotton Technology,
Mumbai (Maharashtra), India

Sushil Yadav
Regional Research Station of ICAR-Central Arid Zone Research Institute, Kukma
Bhuj (Gujarat), India

Jaya Jumrani
ICAR-National Centre for Agricultural Economics and
Police Research, New Delhi, India

Anshuman Singh
ICAR-Indian Agricultural Research Institute, New Delhi, India

SSPH

SATISH SERIAL PUBLISHING HOUSE
403, Express Tower, Commercial Complex
Azadpur, Delhi-110033 (India)
Phone : 011-27672852, Fax : 91-11-27672046
e-mail : info@satishserial.com, hkjain1975@yahoo.com
Website : www.satishserial.com

Published by :

SATISH SERIAL PUBLISHING HOUSE
403, Express Tower, Commercial Complex, Azadpur, Delhi-110033 (INDIA)
Phone : 011-27672852 Fax : 91-11-27672046
E-mail : info@satishserial.com, hkjain1975@yahoo.com

© Publisher

ISBN 978-93-86200-87-7

© 2019. All rights reserved, no part of this publication may be reproduced, stored in a retrieval system or transmitted in any form or by any means, electronic, mechanical, photocopying, recording or otherwise without the prior written permission of the publisher and also the copyright, rights of the printing, publishing , e-book of this edition and subsequent editions will vest with the publisher. All Computer floppies, CD's, e-book and in any other form relating to this book will be exclusive property of the publisher.

This book contains information obtained from authentic and highly regarded sources. Reasonable efforts have been made to publish reliable data and information, but the publisher cannot assume responsibility for the validity of all materials or the consequences of their use. The publisher have attempted to trace and acknowledge the copyright holders of all material reproduced in this publication and apologize to copyright holders if permission and acknowledgements to publish in this form have not been obtained. If any copyright material has not been acknowledged please write and let us know so that we may rectify it.

Composed, Designed & Printed in India

Acknowledgement

In undertaking this study, we have received help from a wide variety of sources. Our greatest indebtedness is to our Course Co-ordinators Dr. S.K. Nanda, Principal Scientist and Dr. R. Venkata Kumar, Principal Scientist, NAARM, Hyderabad who always took keen interest in each and every activity and ensured its successful completion. We feel deeply grateful to Dr. N. H. Rao, Joint Director, NAARM, Hyderabad for their inspirational motivation. We are also thankful to Dr. K.M. Reddy, FET Committee Chairman. We are deeply indebted to FET Committee, NAARM, Hyderabad for their constant support, inspiration, encouragement and valuable suggestions throughout the training.

Beyond words is our gratitude to Dr. D. Swarup, Director, CIRG; Dr. Ashok Kumar, Principal Scientist and Dr. Vijay Kumar, Scientist, FET Programme Co-ordinators, CIRG, Makhdoom, and Shri Niranjan Prasad for their whole hearted support, assistance and scholarly advice to workout PRA techniques during FET training. We would like to thank all the scientists of CIRG, Makhdoom for their hearty support throughout the period.

We are equally grateful to farmers and villagers of Bar ka Nagla village for their utmost courtesy and friendliness. We owe a special debt to Shri Swarup Singh, a progressive farmer of Bar ka Nagla village for his constant support throughout the village training period. We are also thankful to Shri. R.L. Chaudhary, Assistant Manager, Syndicate Bank.

We would like to thank our friends from NAARM for their unconditional support and memorable team efforts during FOCARS training, which are and will be cherished throughout our ARS careers.

Project Leaders

List of Plates

List of Figures

List of Tables

Contents

Chapter 1

Introduction

India is an agricultural country and one-third of its population depends on the agriculture sector directly or indirectly. Agriculture continues to be the mainstay of our economy. Indian agriculture contributes about 14.2 per cent to the National Gross Domestic Product. With food being the crowning need of the mankind, much emphasis has been on commercializing agricultural production. Hence, adequate production and even distribution of food has lately become a high priority global concern. With the changing agricultural scenario and global competition, there is a need of exploiting the available resources at maximum level.

In the new millennium, the challenges in Indian agricultural sector are quite different from those met in the previous decades. The enormous pressure to produce more food from less land with shrinking natural resources is a tough task for the farmers. To keep up the momentum of growth, a careful economic evaluation of inputs, like seeds, fertilizers, irrigation sources, etc. are of considerable importance.

Considering the irrigation needs in Indian agriculture, much emphasis was given to promote the proven cost-reducing micro-irrigation technology of drip irrigation that helps to conserve water, reduces fertilizer inputs and ensures higher productivity. Farmer awareness programmes coupled with subsidy incentives may prove to be some of the helpful strategies. The sustainable method of irrigation needs to be popularized. Problems like salinity in the commands of major irrigation systems need to be minimized by recognizing and incorporating corrective measures. Apart from this multiplication, distribution and availability of good quality seeds is crucial to accelerated food production. To meet the growing competition, companies should adopt modern processing technologies and seed growers have to be trained in cost reducing methods of growing quality seed material.

Rural women play a vital role and participate in all stages of crop production, like transplanting, manuring and fertilizing, harvesting, threshing, winnowing, drying and carrying the products. Other than this, they were actively participated in other livelihood i.e. goat and buffalo rearing. To strengthen this, we need to train the rural woman.

To mitigate and reduce this wide level problem, we need to study the area specific problem solution. For this type of study, Participatory Rural Appraisal (PRA) is the tool that helps in identification of problem and researchable issues to find out possible solutions for short-and long-term benefits. PRA is a good exposure for the trainee scientists to understand the problems of the farmers. It allows them to think individually in his/her area of interest for necessary intervention. PRA also provides an opportunity to the farmers to analyze their livelihood issues.

PRA is an exercise that deals with temporal and spatial issues in agricultural activities. Spatial information supplies the database about soil type, water resources, land use, land pattern and topography of different life support systems. Temporal data helps in seasonal analysis, adoption and awareness about the new technologies and human dimension helps in knowing the facts about participation, gender issues, disaggregation, cultural values and ethics of the people of that locality.

The village selected for our PRA exercise was Bar ka Nagla, which was located at a distance of 4 kms from Central Institute for Research on Goats (CIRG), Makhdoom. The main source of livelihood of this village was agriculture and livestock.

About the FET Centre – CIRG, Makhdoom

The Central Institute for Research on Goats (CIRG) was established by the Indian Council of Agricultural Research (ICAR), Ministry of Agriculture, and Government of India at Makhdoom village near Farah town, in Mathura district of Uttar Pradesh on July 12, 1979 for undertaking research on all aspects of goat production and utilization. The vision is to develop the poor man's cow - the goat as a source of livelihood security, poverty alleviation and employment generation for the smallholders. And the mission is to enhance and then sustain goat productivity in respect of meat, milk and fibre through Research, Extension and HRD support. The institute is located at almost equal distance between Agra and Mathura. The Institute has facilities to conduct research on Goat Genetics and Breeding, Feed Resource Development & Utilization, Nutrition, Physiology, Reproduction, Management, Diagnosis & Prevention of Diseases, Transfer of Improved Technologies and Milk and Meat Products Technology. The Institute has three Goat Breeding Farms on Jamunapari, Barbari and Jhakrana breeds and one Sheep Breeding Farm on Mujaffarnagari breed for conducting research on Genetics and Breeding, besides three Experimental Flocks for conducting research on Nutrition, Health, Physiology, Reproduction and management aspects. The institute is well-surrounded by natural resources and vegetation, and the main attractions were peacocks, parrots and leopards.

❑❑❑

Chapter 2

Objectives

The overall aim of the FET is to provide ARS probationers an opportunity for gaining first-hand experience and insight into agricultural and rural development scenario, in general and the farming situation in respect of the theme-based enterprises in particular.

The specific objectives of the FET are:

1. To provide an opportunity to Scientist-trainees to interact with the farming community and gain insight into their livelihoods.
2. To study socio-economic implications and consequences of technologies, products and processes.
3. To generate a detailed account of technology products and processes, used by producers with respect to production scenario, problems, opportunities, and futuristic approaches.
4. To inculcate culture of teamwork and multi-disciplinary perspective among scientist-trainees.
5. To provide comprehensive insight into the role of private industry and other stakeholders in the value chain.

❑❑❑

Chapter 3

Participatory Rural Appraisal (PRA) Tools

Participatory Rural Appraisal (PRA) is a systematic, semi-structured activity conducted on site, by a multi-disciplinary team. It is basically a bottom-to-top approach of learning rural life forms with and by the rural people. Due to the active participation of community members as well as scientists, PRA has become a useful method to focus attention on people, their livelihoods and relationship with socio-economic factors. It is a good technique to help the community members make an appraisal of their livelihoods and issues related to it.

Diverse information is collected during PRA, using a number of techniques. This information is verified by triangulation among various key informants (KI's) in addition to on-site observation by the team members. The various techniques used by the FET team to collect information are as follows:

1. Direct observation, collection and verification of primary data.
2. Collection of secondary data from Panchayat members and officials of the State Deptt. of Agriculture.
3. Semi-structured interview by raising some general issues to motivate villagers' participation in different aspects of the village.
4. Collection of detailed views by approaching some representative key informants.
5. Use of snowball technique to know more about some specific issues.
6. Participatory mapping for rapport building, easy interaction and spontaneous involvement.

7. Seasonal analysis to understand different time-bound farm activities, problems and opportunities throughout the year.
8. Problem identification and preference ranking to identify the top-most problems and preparing the solution plan accordingly.

❑❑❑

Chapter 4

FET Methodology

The FET is a learning process. It is firmly anchored on the principle of 'seeing is believing' and 'learning by doing'. FET has three phases. In the first phase, the village for PRA work was selected in consultation with the FET co-ordinator, Dr. Ashok Kumar and Dr. Vijay Kumar, Program Co-ordinators, CIRG, Makhdoom, Mathura (UP). Our study village Bar ka Nagla was located at a distance of 4 km from CIRG, Makhdoom. In Bar ka Nagla, we chose three main key informants (KIs), Shri Nawab Singh Bhagel, Shri Tikam Chaudhary and Shri Vasudev Singh for our transect walk. The team walked and interacted with the KIs and villagers, observed while walking and noted the information. Interaction with KIs was initiated from the semi-structured questionnaires and then branched as per the need. The information given by one KI was triangulated (verified) with other KIs and on the spot analysis. The principle and methods of PRA was applied for gathering information on 22 PRA techniques.

In the second phase of FET, we visited and interacted with various stakeholders, like the scientists from KVK, Block Development Officer, Panchayat Officer, Veterinary Officer and several village administrative officers. The interaction with stakeholders also helped to understand the agricultural problems and prospects in the village and to prioritize the research problems for action. To understand the industrial scenario prevailing in the area, we visited Private Cold Storage near the village.

In the third phase, we shared our experience with the villagers and also with the scientists of KVK, Mathura by organizing village seminar on 21 November, 2011 and institute seminar at CIRG,

Makhdoom on 24 November, 2011, respectively. We obtained feedback from the farmers and the scientists, in the light of which the action plan was modified and proposed.

□□□

Chapter 5

PRA Techniques used in the Village

The techniques used in the village for PRA survey was given below:

1. Basic information about the village
2. Transect walk
3. Agro-ecological map
4. Resource map
5. Seasonal calendar (activities)
6. Seasonal analysis (problems)
7. Gender disaggregated activities
8. Social map
9. Time line
10. Time trend
11. Mobility map
12. Venn diagram
13. Bio-resource flow
14. Wealth ranking
15. Livelihood analysis
16. Daily activity and routine
17. Indigenous technical knowledge
18. Technology table and technology map
19. Matrix ranking
20. Consequence diagram
21. Problem identification and ranking
22. Problem tree and solution tree

5.1 Basic Information

Key Informants	:	**Swarup Singh, Jaswant Singh, Kunwar Singh**
Name of the Village	:	**Bar ka Nagla**
Village Panchayat	:	**Visu Gram Panchayat**
Mandal	:	**Agra**
Taluka	:	**Mathura**
District	:	**Mathura**
State	:	**Uttar Pradesh**
Climatological Data		
Mean Annual Rainfall	:	**345.6 mm**
Mean Annual Temperature	:	**32.9 ^{0}C**
Relative Humidity	:	**46 %**
Topography	:	**Plain**
Geographical Co-ordinates		
Latitude	:	**27^0 18′ 15.25" N**
Longitude	:	**77^0 45′ 35.27" E**
Altitude	:	**576 ft MSL**
Geographical Area		
Total Area	:	**50 ha (200 bigha)**
Gross Cropped area	:	**44 ha**
Net Cultivated Area	:	**40 ha**
Irrigated Area	:	**40 ha**
Rainfed Area	:	**Nil**
No. of Tanks	:	**Nil**
Cropping Intensity	:	**110%**
Residential Area	:	**4 ha**
Demographic Pattern		
No. of Households	:	**76**
Total Population	:	**611**
No. of Males	:	**326**
No. of Females	:	**285**
Literacy	:	**73.9%**

Community Categorization		
Total Population	:	**611**
No. of Hindu Population	:	**586**
General	:	**147**
OBC	:	**344**
ST	:	**57**
Muslim	:	**25**
Occupational Distribution of Families	:	
No. of Families involved in Agriculture	:	**61**
No. of People involved in Agriculture	:	**240**

Labour includes agricultural labour and non-agricultural labour including driving, tailoring and shop-keeping.

Agricultural Scenario

A. Land Distribution		
Medium (5-10 ha)	:	**10**
Small (2-5 ha)	:	**157**
Marginal (<2 ha)	:	**05**
B. Crop Husbandry		
Kharif season	:	Paddy, Dhaincha, Bajra, Sorghum, Seasamum
Pre-Kharif (summer) season	:	Sorghum
Rabi season	:	Mustard, Potato, Wheat
C. Animal Husbandry		
Type of Animals	:	Buffalo, Cow, Sheep, Goat, Pig and Poultry

Animal	No of Population	Breed
Cow	20	Jersey, Local
Buffalo	100	Murrah, Jaipuri, Local
Sheep	25	Faki
Goat	130	Jamunapari, Jakhrana, Barbari and Local
Donkey	5	Local
Mule	1	Hybrid
Backyard Poultry	25	Desi

Mortality Rate of Different Animals: Buffalo, Goat, Donkey, Cow and Poultry due to natural mortality mainly, disease mortality is rare.

D. Water Bodies

Open well : **03**

Bore well : **43**

Hand Pumps : **17**

Pond : **02**

E. Others

Soil Type : Predominantly Domat, Baluee and Pila soils prevail.

F. Agricultural Implements

Tractor : **05**

Tractor trailer : **02**

Cultivator : **04**

Power Sprayer : **03**

Seed-cum-fertilizer drill : **02**

Disk harrow : **04**

Potato planter : **03**

Potato digger : **03**

Land leveller : **04**

Chaff cutter : **50**

Plate 1. A satellite image of Bar ka Nagla Village (*Courtesy :* Google Earth, 2009)

5.2 General Transect

Key informants : Swarup Singh & Jaswant Singh

Transect is the walk through the village in a particular direction along with some key informants. The main objective of the transect is to understand and study the major land uses, topographical pattern, water resources, natural vegetation and different ecological zones by observing, interacting and discussing with the key informants, while walking in the decided direction. The items for discussion included topics, such as land use pattern, land slope, soil type, soil fertility status, water resources, agro-ecological zones, crops, weeds, trees, livestock and institutions.

The entire village of Bar ka Nagla was divided into three zones on the basis of soil type, *viz. Domat, Baluee and Pila mitti*. The residential area was towards the central part of the village with agricultural fields, lying towards south and east direction of the village. The soil of this village was predominantly Damod, Baluee and Pila. The main crops were Potato, Wheat, Mustard, Bajra and Sesamum.

Like other villages, the farming practise is not a traditional one i.e. not with animal drawn implements, like country plough, rather it is more mechanized and generally uses tractor with cultivator and land leveller for land preparation; seed-cum-fertilizer drill and potato planter for sowing; sprayer for spraying pesticides; chaff cutter for cutting straw; harvester and potato digger for harvesting. The livestock population of the village consists of cows, buffaloes and goats.

Plate 2. Transect walk with the villager

Bar ka Nagla village was situated near Farah town, so they don't have enough facilities, like schools, hospitals or health centre. They visit Farah town for their daily activities, such as for School, Colleges, Medical care, Maket, etc. Still then the village have one Anganwadi and four temples, as the majority of population were Hindu. Saline water, electricity, death of goats due to diarrohea, FMD in cattle, parasitic infestation in buffalo, frost injury in potato, aphids and white rust in mustard, termites in wheat, unsafe chaff-cutter, lack of proper extension activities, lack of new technology adoption, lack of awareness for soil testing, unavailability of primary health centre and veterinary hospital, etc. Proper vaccination and de-worming for goats and cattle, goat farming for improving rural women empowerment, formation of need based SHGs for farmers, utilization of land for horticultural crops, irrigation by using drip, micro and sprinkler system, good rapport building with KVK, introducing apiculture & poultry farming, training and awareness for proper heath care and managemental practices for livestock and agricultural implements, etc., were identified as the major opportunities for the village.

Plate 3. Rapport building with the villagers and children

Plate 4. Discussion with the villagers about their agricultural livelihoods

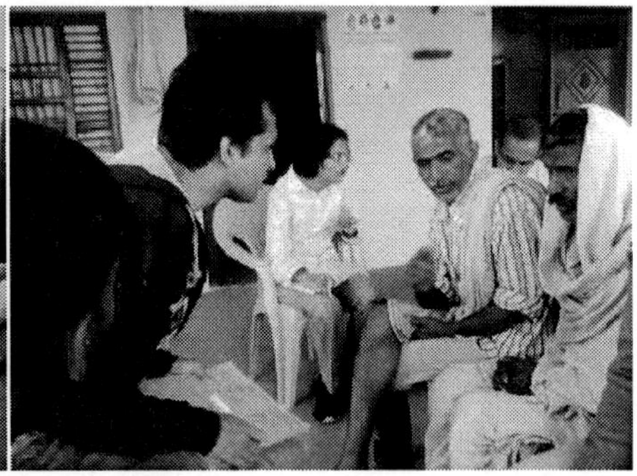

Table 1. The general transect of Bar ka Nagla village

S.No.	Particulars	Zone 1	Zone 2	Zone 3
1	**Soil Type:**	Domat (Loamy)	Baluee (Sandy loam)	Pila (Yellow soil)
2	**Major Crops**	Potato, wheat, mustard, bajra, daincha	Potato, mustard	Mustard
3	**Livestock/ Poultry**	Buffalo, Cow, Goat, Piggery, Poultry	Buffalo, Cow, Goat, Piggery, Poultry	-
4	**Trees**	Neem, Shisham, Ber and Banyan	Neem, Shisham, Ber and Banyan	Desi babul (Acacia nilotica) and bilayti babul (Prosopis julifera)
5	**Weeds**	Gumma (Achyranthes aspera), Satyanasi (Argemone maxicana), Jungli jo (Avena fatua), Gehu ka mama (Phalaris minor), Bhathua (Chenopodium album), Krishna neel (Commonella (Amranthus)	Bhathua (Chenopodium album), Krishna neel (Commonella bengalenssis), Hazar dana (Phyllanthus niruri), Katra (Calotropis gigantia), Jungli palak (Purslane poleracea), Jungli	Katra (Calotropis gigantia), Jungli palak (Purslane poleracea), Jungli gobhi (Brassica oleracea), Cholaie

		bengalenssis), dana (Phyllanthus niruri), Katra (Calotropis giga-ntia), Jungli palak (Purslane poleracea), Jungli gobhi (Brassica oleracea), Cholaie (Amranthus)	Hazar gobhi (Brassica oleracea), Cholaie (Amranthus)	
6	**Water Sources**	Bore well	Borewell	Rainfed
7	**Insects**	Termite, Kernal blunt, Early shoot borer, Late blight, leaf roll virus, potato aphids, Aphids and white rust	Late blight, leaf roll virus, potato aphids, Aphids and white rust	Aphids and white rust
8	**Diseases**	Late blight, leaf roll virus, and white rust	Late blight, leaf roll virus, and white rust	White rust
9	**Diseases in livestock**	FMD, Diarrohea, Bloat, Parasitic infestation	FMD, Diarrohea, Bloat, Parasitic infestation	-
9	**Problems**	Saline water, Electricity, death of goats due to diarrohea, FMD in Cattle, Parasitic infestation in buffalo, Frost injury in potato, Aphids and white rust in mustard, Termites in wheat, Unsafe chaff-cutter, Lack of proper extension activities, Lack of new technology adoption, Lack of awareness for Soil testing, Unavailability of primary health centre and veterinary hospital, etc.		
10	**Opportu-nities**	Proper vaccination & de-worming for goats and cattle, Goat farming for improving rural women empowerment, Formation of need based SHGs for farmers, Utilization of land for horticultural crops, Irrigation by using drip, micro and sprinkler system, Good rapport building with KVK, Introducing apiculture & poultry farming, Training and awareness for proper heath care and mangemental practices for livestock, etc.		

5.3 Agro-Ecological Map

Key informants: Jaswant Singh and Hardayal

Agro-ecological map indicates the macro and micro ecological (subsystem) features in a village. The meteorological parameters, like rainfall, temperature, relative humidity flora and fauna of the village and the basic land use pattern, such as agro-forestry, forest cover and wasteland is depicted in the map. The village Bar ka Nagla was located at 2 km towards west of Farah town on NH-2 Grand Trunk Road. It was situated at latitude 27^0 15″N, longitude 77^0 35″ E and altitude of 576 ft above sea level. The area was characterized by hot summer and cold winter and comes under irrigated area. The soil type was predominantly Domat, Baluee and Pila soils prevail in the village.

The average rainfall was 345.6 mm in 49 wet days. The rainfall is uneven in distribution and highly intense in nature, causing runoff and erosion problem. Maximum rainfall is distributed in July, August and September months. The average temperature recorded in 2010-11 was 32.9^0C, the maximum being 48^0C with minimum of 2^0C. December and January are the coldest months, while May and June are the hottest. Maximum crop area is under rabi crop, where 100% cultivable area is under crops, followed by Kharif with 50% area under crop cultivation. In Summer (Zaid), 10-15% cultivable area is under crop cover. The cropping intensity is 110 per cent.

In Rabi season Mustard, Wheat, Potato and Berseem occupies the major area. Potato 3797, Kufri Sutlez, Spl-1 are popular varieties adopted by farmers. Rohini, Type-69 and Chipka are widely cultivated mustard varieties in the village. While, varieties, like Lok1, 2009, & HD 2329 predominate area under wheat cultivation. In Kharif season Dhaincha and Bajra are cultivated. Sorghum variety MP Cherry was cultivated for fodder purpose with multiple harvesting at interval of 25-30 days. No pulse crops were cultivated in the villages due to high salinity in soil and water, as pulses were highly sensitive to salinity.

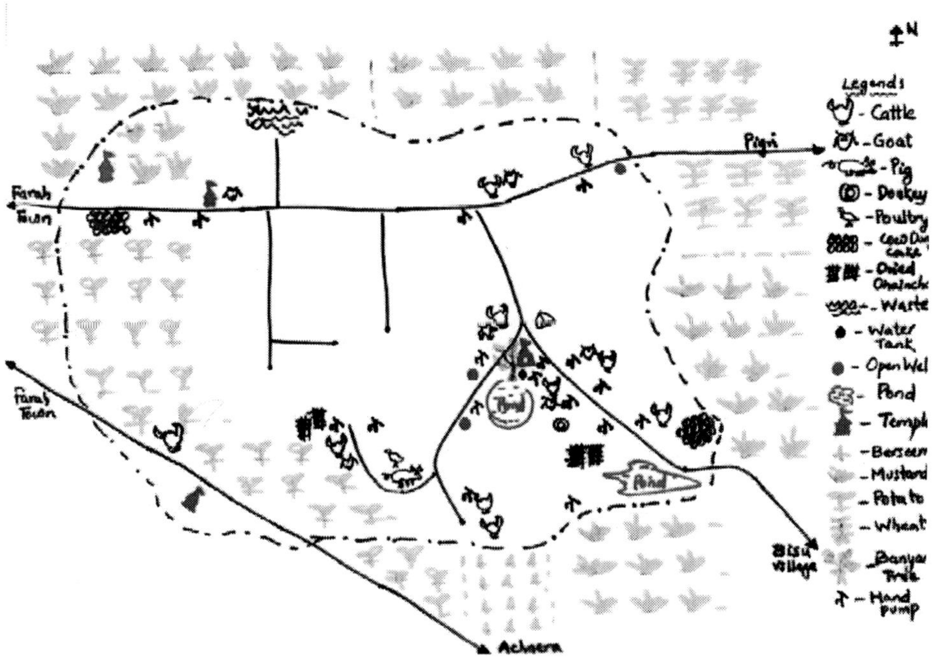

Figure 1. Agro-ecological map of Bar ka Nagla village, depicting the various agricultural and ecological resources of the village

Animal husbandry was diversified in the village with goat and cattle farming, sheep rearing and piggery; Murrah, Jaipuri, Local breed in Buffalo; Jamunapari, Jakhrana and local breeds in goats; and in cow local & jersey breeds.

Tropical thorny trees Acacia nilotica and Prosopis julifera are widely distributed in the ecosystem. Trees are grown as the boundary of fields, and also near the residential areas, the main species being Neem, Subabul, Babul, Banyan and Ber. The major weeds are Wild rice (*Echinocloa colonum & Echinocloa crusgali*), Badha (*Cyperus defformis*), Gumma (*Achyranthes aspera*), Satyanasi (*Argemone maxicana*), Jungli jo (Avena fatua), Gehu ka mama (*Phalaris minor*), Bhathua (*Chenopodium album*), Krishna neel (*Commonella bengalenssis*), Hazar dana (*Phyllanthus niruri*), Katra (*Calotropis gigantia*), Jungli palak (*Purslane poleracea*), Jungli gobhi (*Brassica oleracea*), etc.

Plate 5. The weed diversity of Bar ka Nagla village (Cholaie & Katra)

5.4 Social Map

Key informants : Meera Sharma and Jaswant Singh

The social map gives the information to understand and analyze the social structure, stratification, social institutions and availability of social facilities in the context of existing socio-economic condition of the village. According to key informants, the village have population of 611 living in 76 houses, out of which 326 were males and 285 were females. The total area of the village is 200 bigha i.e. 50 ha. The main occupation of the villagers were agriculture and animal husbandry. The villagers were also engaged in professions, like teaching, truck driving, cycle rikshaw drawing, sewing, shop keeping, etc. Some of the notable social features were given below.

Social structure and stratification : The predominant religion in the village was Hindu. Few Muslims were located in spatially isolated place in the village. Pandits and Baghels are predominant castes in the village. The population of other backward class (OBC) comprising of Baghel, Jats, Yadav and Mali, followed by Pundits (Brahmins) and Thakurs (General). Dhobi & Mahal (Scheduled Tribe) dwelling places are distributed in pockets between residential areas of other castes. All the castes have equal access to resources in the village. About 96 per cent of the village population belongs to Hindu religion.

Pandits and Thakurs were predominant castes. Leadership pattern was based on economic and caste criteria. The Hindus were vegetarians; consumption and cooking of meat was banned in the village. The Muslims were isolated from mainstream society.

Housing pattern : Most houses were Pucca with RCC roof. Few thatched Kaccha houses were also distributed irregularly in the village.

Most of the families were with small and medium land holdings. Joint families dominate in the village and only few families were nuclear in nature.

Plate 6. Socially important institutions in the village Bar ka

Government institutions : No government institutions were found in the village except one Anganwadi. The children were used to go nearby Bisu village for primary education, where also the gram panchayat office is located. About 73.9 per cent of population were literate in the village, which was more than literacy rate of Farha block and Mathura district.

Plate 7. Religious institutions in Bar ka Nagla

Local institution : There were 4 temples in the village, namely Rathwari maiah temple, Lord Hanuman temple, Bholebaba temple and Balaji temple. Diwali and holi were prominent festivals celebrated in the village with inter faith and harmony among different castes.

Other institutions : An NGO named Mahila Samakhya from Mathura has organized two women Self Help Groups in the village, each with 20 members. The SHGs were meant for producing and marketing processed food products, like pickles, thus generating money for the group. Also, an NGO, named Surya foundation was providing free education to the School dropouts.

Social evils : According to key informants, early marriage (14-16 years) is the major social evil in the village. The village is relatively free from alcoholism, child labour, dowry and social conflicts.

5.5 Resource Map

Key informants : Vinod and Ramprakash

The village resource map shows the different kinds of natural and man-made resources, such as land, vegetation, common land use, soil, livestock, transport and source of communication. During the general transect through the village, the different resources found are listed below :

Land	:	Total land area of this village is 50 ha (200 Bigha), out of which 40 ha was cultivated.
Soil	:	The soil type was predominantly sandy loam and saline-sodic in nature.
Pond	:	Two ponds existing in the village were being utilized to provide drinking water for livestocks.
Govt. organization	:	One Anganwadi
Crops	:	Mustard, Potato, Wheat, Bajra, Dhaincha, Sesamum
Livestock	:	Buffalo, Goat, Sheep, Pig, Donkey, Mule and Backyard poultry.

Crop cultivation in the village was highly mechanized, with farmers having their own farm machineries, like tractors, cultivators, seed drill, potato planters, power sprayers, etc. Farm machineries were also hired on rent basis from surrounding villages and Farah town. The irrigation source was mainly done with borewells with the help of diesel engine. Livestock management in the form of vaccinations and artificial insemination were not so prevalent in the village.

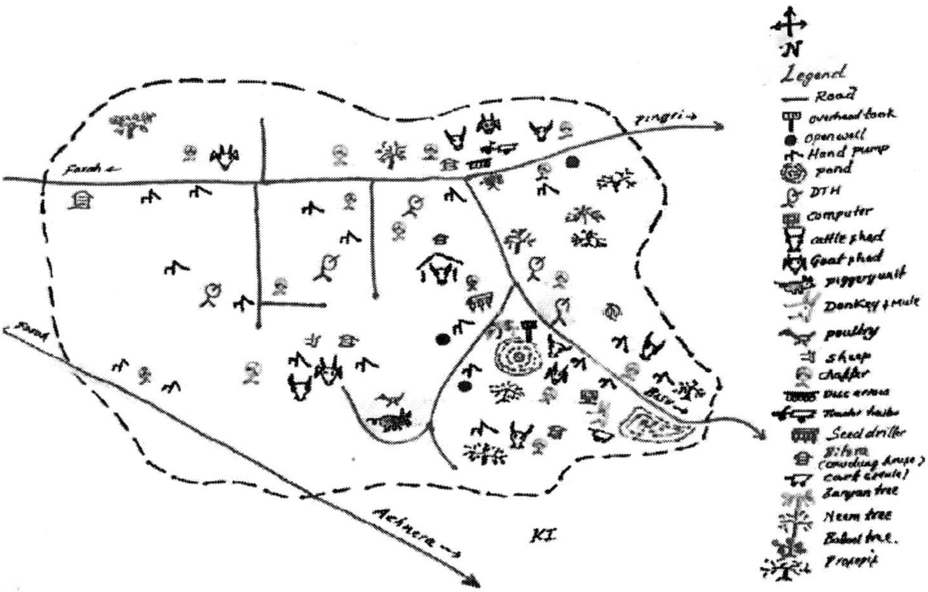

Figure 2. Resource Map of Bar ka Nagla village

Table 2. Resource of Bar ka Nagla village

Particulars	Articles
Transport facilities	Auto rickshaw, Motor bike, Bicycle and Van
Communication facilities	Television, Radio, Mobile, DTH.
Supply and service agency	Cold storage, General Store, Syndicate bank (Faraha)
Agriculture implements	Tractor, Tractor trailer, Cultivator, Power sprayer, Seed cum fertilizer drill, Disk harrow, Potato planter, Potato digger, Land leveller, Chaff cutter and Power sprayer
Animals use for agriculture	Complete Mechanization
Advisory facilities information	Central Institute for Research on Goats (CIRG), Makhdoom, Krishi Vigyan Kendra, Mathura, Mahila Samakya NGOs, Surya foundation, Pandit Deendayaldham NGO (Nagla Chandrabhan), Veterinary Hospital (Faraha), Agriculture office (Farah).
Animal Reproduction	Natural service, no Artificial Insemination (AI) due to lack of awareness

5.6 Venn Diagram

Key informant : Chandan, Mohan Singh and Virender Singh

The Venn diagram shows existing institutions, organizations, groups and important individuals found in the village as well as the villagers' view of their importance in the community. Venn diagram also indicates the level of contact and co-operation between these organizations and groups. The organizations/institutions/groups around village are listed below in order of their importance:

- Central Institute for Research on Goats, (CIRG, Makhdoom): Provides technological interventions in the form of training, providing animal breeds and health care services, like vaccination.
- Agriculture Department, Farah: Provides crop related information, seed, fertilizer & other inputs under various developmental programmes.
- Syndicate Bank, Farah is the only institutional source of finance to farmers in the form of crop and animal husbandry loans. Thus, bank influence the micro-enterprise in the village.
- Pandit Dheendayal Dham, an NGO at Nagla Chandrabhan imparts vocational training to rural youths. Its Hospital and community hall services are utilized by the villagers.
- Input dealers of Farah were source of seeds, fertilizers & pesticides for the village. They are informal source of crop cultivation information for farmers.

With respect to potato production, the organizations/institutions/groups around Bar ka Nagla village are listed below in order of their importance:

- Cold storage Unit, Bisu: Farmers store harvested potatoes in cold storage unit on rent basis. This helps them to fetch remunerative price for their produce, when prices are stabilized in market.

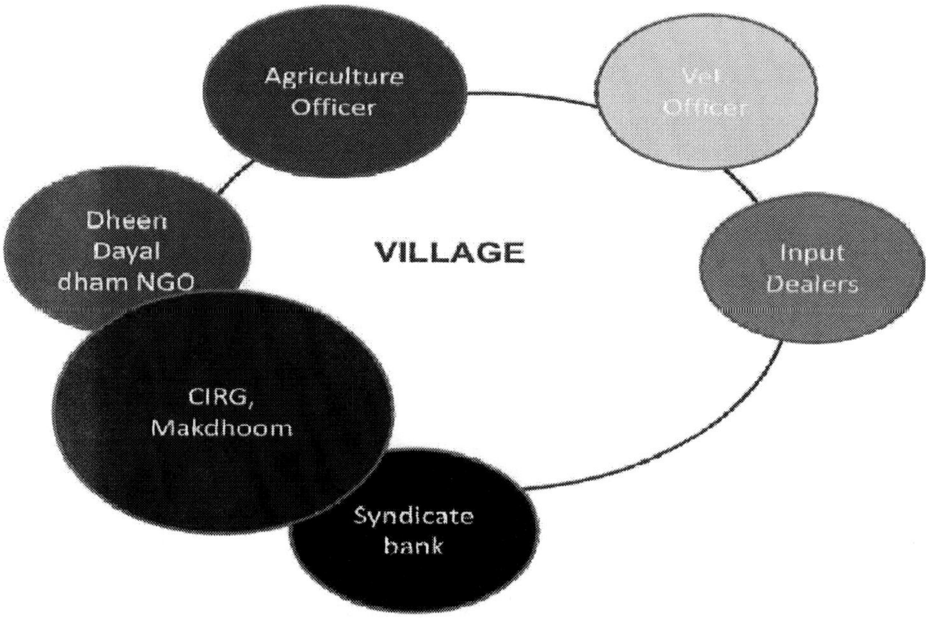

Figure 3. Venn diagram of agricultural activities in Bar ka Nagla

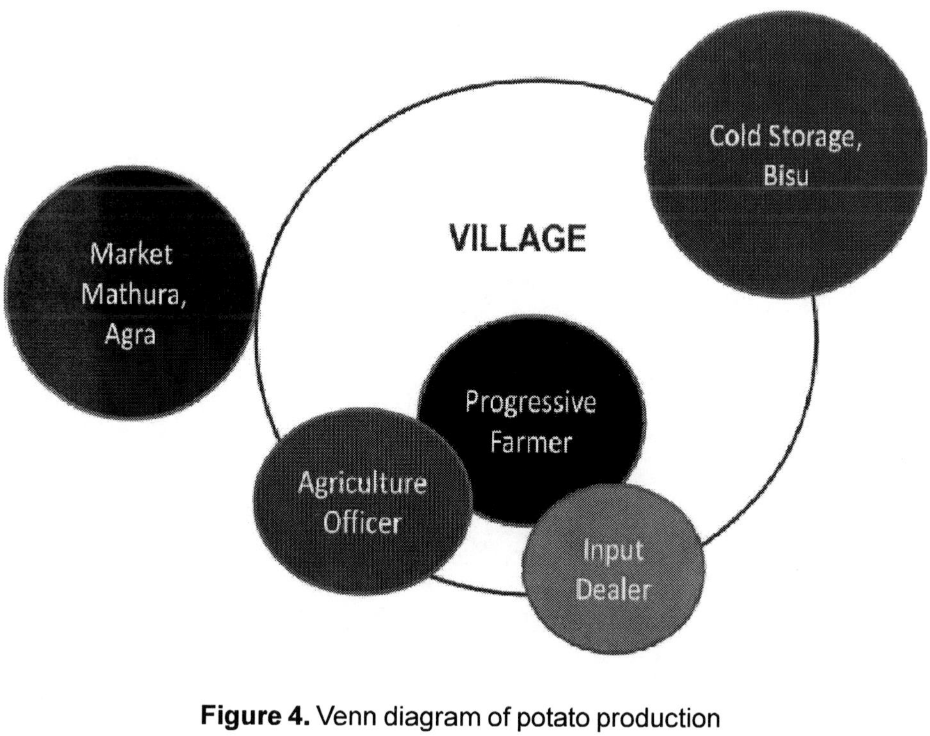

Figure 4. Venn diagram of potato production

- APMC, Mathura and Agra are the nearby markets, assured market influenced potato cultivation in the area.
- Progressive farmers: Own potato digger and planter, which are resources for other potato growers, who hire on rent basis.
- Agriculture department (RSK), Farah: Provides advisory facilities to farmers on potato cultivation aspects.

5.7 Bio-Resource Flow

Key informants : Amar Singh and Hardayal

Bio-resource flow diagram reflects the inflow and outflow of farm produce and its by-products from and to the household. It explains the inter-relationship between different farm enterprises that enable holistic planning for development of farm household. In the village Bar ka Nagla, the typical representative household is small-sized farmer. The existing practices are shown by solid black lines. Proposed interventions are depicted by dashed red lines.

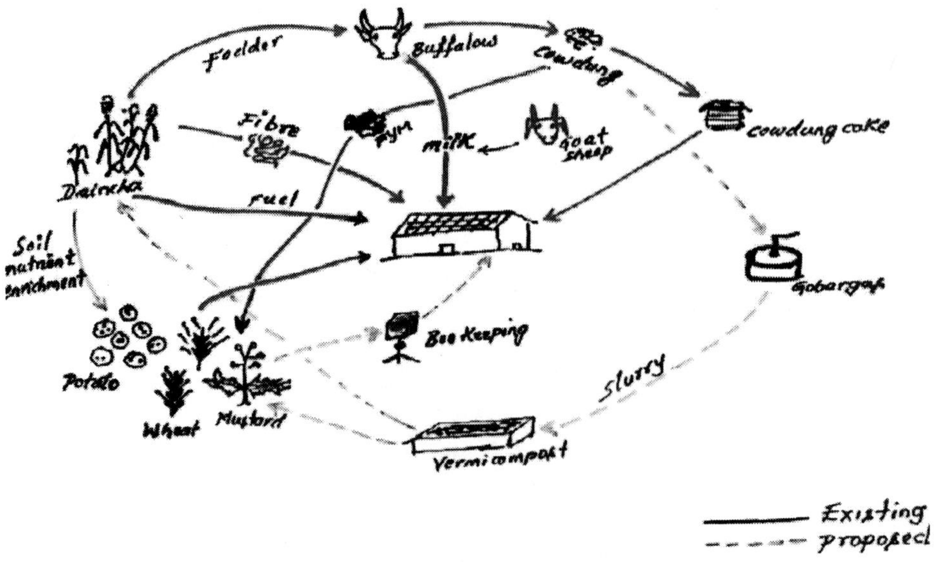

Figure 5. Bio-resource flow diagram

5.8 Indigenous Technical Knowledge (ITK)

Indigenous technical knowledge is the knowledge, or practice possessed by the villagers, which are passed on from generation to generation by word of mouth. ITK bank is very helpful to the researchers in collecting the vast experiences and long experiments by the villagers, improving their level of interactions and demonstrating these practices to the outsiders. The following (Table) shows some of the Indigenous Technical knowledge provided by the villagers of Bar ka Nagla.

Table 3. ITKs in Bar ka Nagla village

S.No.	Practice	Purpose	Process	Location	KI
Agriculture					
1	Ash application	Aphidicide	Ash (1kg) and potash (50 g) broad casting on crop	Anganwadi	Raviprakash
2	Neem cake/oil application	Termite control	Neem cake/ oil is added to soil	Banyan tree	Hardayal
Animal Husbandry					
3	Feeding jeera, fenugreek , hing and gur paste to animals	To relieve bloating	Jeera (10g), + Fenugreek (10g) + Hing (5g) + Gur (250 g)	Bholebaba temple	Mohan Singh
4	Methi-Mustard oil paste application in animals	Foot and Mouth Disease	Mustard oil (250 ml) oil+methi (250g) + Gur (500 g) Paste application on affected parts	General store	Virender Singh
5	Khatra weed feeding	To enhance cow milk yield	Entire herb is given as feed	Mustard field, pond	Amar Singh
6	Marigold leaf extract application	To treat eye pain in animals	Leaf extract drops is applied to eyes	Banyan tree	Ramprakash

7	Bael leaf extract Swabbing in cattle.	To reduce insect pest infestation cloth on skin of animals	Leaf extract is swabbed with cotton	Bholebaba temple	Hardayal
8	Gobhi feeding	To treat constipat-ion	Green leaf of cauliflower and cabbage fed to cattle.	Anganwadi	Rani
Medicine					
9	Guar leaf	Wound healing	Guar leaf extract application on cut wounds	Anganwadi	Nawab Singh Baghel
10	Balan *(Abutilon indica)*	Arthritis	Oil extract from leaves	Mahila Samakya SHG	Meera Sharma
Others					
11	Goat dung and usar soil as detergent	To remove stains in clothes	Clothes are dipped in water contai-ning goat dung and usar soil and slightly heated	Hanuman temple	Jaswant Singh

5.9 Daily Activity and Routine

Daily activity relates to the economic activity of the village youth and middle-aged men and women. Contrary to this, daily routine includes both economic and general activities. In order to have precise knowledge regarding daily activity and routine, the key informants must be middle-aged men or women, belonging to farming community. The main purpose of these PRA tools is to sort out proper time to intervene with various extension activities. However, our target was to find out proper time for interviewing farmers to acquire knowledge on various aspects of social life as well as discussions for organizing village seminar. The daily activities and routine are depicted in Table 4, 5, 6, and 7.

Table 4. Daily activity profile of Middle-aged men

Key informants : Bhoopinder Singh and Uday Singh

Time	Activity
6:00 AM to 7:00 AM	Feeding of Livestock & milk collection
7:30 AM	Milk delivery to villagers
7:30 PM	Milk delivery to Tirumala broker
9:00AM to 12:30PM	Farm activities
1:00PM to 5:00 PM	Farm activities
5.30PM	Green fodder collection and returning from field to home
7:30 PM	Informal discussion on farming and animal husbandry

Table 5. Daily activity profile of Middle-aged women

Key informants : Mrs. Rupanjya, Mrs. Lilavati and Mrs. Mahadevi

Time	Activity
5:00 AM to 6:00 AM	Green fodder collection
7:00 AM to 11:30 AM	Farm activities and fodder collection
2:00 PM to 5:00 PM	Field activities
5:00 PM to 6:00 PM	Collection of milk

Table 6. Daily routine table for Middle-aged men

Key informants : Mr. Bhoopinder Singh and Mr. Uday Singh

Time	Activity
5:00 AM	Wake up, Refreshing & bath
6:00 AM	Animal feeding and care
6:30 AM	Milking of cattle
7:30 PM	Milk sale to local vendors/halwais
8:30 AM	Breakfast and going to farm
9:00 AM to 12:00 Noon	Farm activity
12.30 PM	Lunch in the field
1:00 PM to 4:00 PM	Farm activity
4:00 PM to 5:00 PM	Fodder collection

5:30 PM	Return from field to home
6:00 PM	Refreshment and visit to local market Farah
7:00 PM	Entertainment (Watching TV), discussion and gossiping with peers & family members
8:00 PM	Dinner
9:00 PM	Go to bed

Table 7. Daily routine table for Middle-aged women

Key informants : Mrs. Rupanjya, Mrs. Lilavati and Mrs. Mahadevi

Time	Activity
4:00 AM	Wake up
4:00 AM to 5:00 AM	Refreshing and Bath
5:00 AM to 6:00 AM	Drinking water to animals and green fodder collection
6:00 AM to 7:00 AM	Breakfast making and eating
7:00 AM to 11:30 AM	Field work
12:00 Noon to 2:00 PM	Lunch preparation and lunch in the field with family members
2:00 PM to 5:00 PM	Field Activity
5:00 PM to 6:00 PM	Milk collection
6:00 PM to 7:00 PM	Dinner preparation
7:00 PM to 7:30 PM	Watching TV, gossiping with family members
8:00 PM	Dinner
9:00 PM	Going to bed

Table 8. Daily routine table for Young men

Key informants : Anil Kumar, Nand Kishore and Mr. Vishnu Kumar

Time	Activity
6:00 AM to 7:00 AM	Wake up, Refreshing & bath
7:00 AM to 7:30 AM	Breakfast
8:00 AM	Go to school
12:00 Noon	Lunch at school
3:30 PM	Coming back to home from school

4:00 PM	Light food
4:30 PM to 6:30 PM	Self-study
6:30 PM	Jogging and playing with friends
7:00 PM to 8:00 PM	Watching TV, Talking with parents
8:00 PM	Dinner
8:30 PM	Gossiping with family members
8:30 PM to 9:30 PM	Self-study
9:30 PM	Go to bed

Table 9. Daily routine table for young women

Key informants : Usha, Laxmi and Ruchi

Time	Activity
5:30 AM	Wake up & quick refreshing
5:45 AM to 6:30 AM	Animal feeding, cleaning of animal and milking
6:30 AM to 7:00 AM	Filling of water
7:00 AM to 7:30 AM	Breakfast making and eating
8:00 AM to 8.30 AM	House cleaning and food preparation
8:30 AM to 9:00 AM	Bathing and washing cloth
9:30 PM to 2:00 PM	College
12:00 Noon to 2:00 PM	Lunch making and serve it to children and in the field
2:00 PM. to 2:30PM	Coming home from college and taking lunch
2:00 PM to 5:00 PM	Field Activity
2:30 PM to 3:00 PM	Watching TV/ Rest
3:00 PM to 5:00 PM	House cleaning and collection of drinking water
5:00 PM to 8:00 PM	Food preparation
8:00 PM to 8:30 PM	Dinner
8:30 PM to 9:30 PM	Self-study
10:00 PM	Go to bed

Table 10. Daily routine table for Aged women

Key informants: Omvati, Mangala

Time	Activity
4:30 AM	Wake up & refreshing
4:35 AM to 5:00 AM	Bath
5:00 AM to 5:30 AM	Feeding animals
5:30 AM to 6:00 AM	Milking
6:00 AM to 6:30 AM	Daliya preparation
6:30 AM to 7:00 AM	Breakfast making and eating
7:00 AM to 8:30 AM	Upla making
8:30 AM to 10:00 AM	Taking care of Grand children
10:00 AM to 11:00 AM	Cloth washing
11:00 AM. to 12:00 Noon	Collect water for family and animals
12:30 PM to 1:00 PM	Lunch
1:30 PM to 2:30 PM	Collection of grass & cutting grass with machine
3:00 PM to 5:00 PM	Feeding animals and watering
5:00 PM to 6:30 PM	Milking and tyeing
7:00 PM to 8:00 PM	Food preparation & eating
8:30 PM	Go to bed

5.10 Mobility Map

Key informants : Mr. Deewan Singh, Mr. Kailash, Mr. Narayan Singh, Mr. Chandan

Mobility map, as an extension tool, provides knowledge regarding day-in and day-out movement of villagers for different purpose. This map is helpful in exploring the spatial mobility of the villagers for the different activities, like health/medical services, marketing, education and entertainment. The KI's were asked to give information on mobility of villagers and information were collected on the basis of location, purpose, distance, mode of transport and transport cost to each location. It was observed that Bisu was the nearest (0.5 Km) and most frequently visited place, basically for social networking, meeting friends and well-wishers. The village children went to Bisu for attending school. There is also Gram Panchayat Office (GPO) in Bisu. The next most visited place was Farah, which was the nearest market (1.5 km)

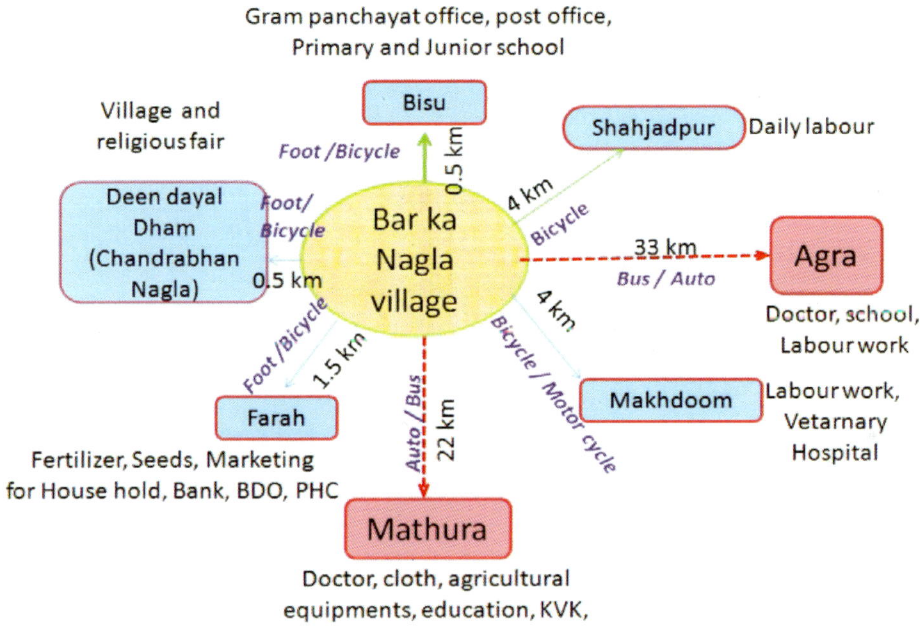

Figure 6. Mobility map of Bar ka Nagla village

from the village. They visited Farah for purchasing household articles, medicines and agricultural inputs, such as seeds, fertilizers and pesticides. Again, Farah town houses the Block Development Office, Primary Health Centre and Nationalized Banks (SBI and Syndicate bank). For treatment of ill goats, farmers visited CIRG, Makhdoom (usually fortnightly). The landless labourers of the village frequently visited Shahjadpur for securing labour work. They used to visit Agra and Mathura for education of children, labour work, consulting doctor, seeing religious and historical places, and purchase of heavy farm machinery. Visits to Mathura and Agra cities were occasional and in case of urgency (say medical purpose).

Pt. Deendayal Dham (0.5 Km from the village) lies on the way to Farah. The purpose of visiting Deendayal Dham was to attend religious fair and to meet peers. Again, some young girls visited Deendayal Dham for attending training in handicrafts. For agricultural problems, like pest, diseases and soils, they made occasional visits to KVK and Veterinary University, Mathura and which is only 22 Km from the village. The main mode of transport was found to be bicycle and motorcycle. Besides, they also visited by auto and bus. Detailed mobility map of Bar ka Nagla village has been given in the Table 11 and Figure 6.

Table 11. Mobility Table for Bar ka Nagla Village

S. No	Place (km)	Distance	Mode	Frequency	Cost (Rs)	Purpose
1	Farah	1.5	Walk/ Bicycle,	Daily	-	Marketing for daily needs, Farm inputs, Banking facility, BDO, PHC
2	Bisu	0.5	Walk,	Daily	-	To meet Bicycle friends, peers and well-wishers, Gram Panchayat Office, Primary and Junior high school
3	Mathura	22	Auto/	Weekly	15	Doctor, clothes, Bus Agri-Equipments, Education purpose
4	Agra	33	Bus/ Auto	Weekly (Auto) 3 (Govt. Bus)	15	Doctor, school, Labour work.
5	Baldev	20	Bus / motor cycle	Monthly	30	Pilgrimage, fair
6	CIRG, Makh-doom	4	Bicycle, Motor cycle	Fortnightly	20	Labour work Veterinary Hospital
7	Sahajad-pur	4	Bicycle	Daily	-	Labour work
8	Deendyal Dham	0.5	Foot/ Bicycle	Frequently	-	Village fair, Religious purpose, Handicraft training, ayurved hospital

5.11 Wealth Ranking

Key informants : Deewan Singh, Vasudev Narayan Singh, Ganesh and Uday Singh

Plate 8. Farmers wealth ranking of Bar ka Nagla

The Gram Panchayat Office of the selected village (Bar Ka Nagla) had detailed records about the socio-economic status, land holding, number of livestock, *etc.* However, wealth ranking was carried out to realize the relative wealth of households in the village, as perceived by the villagers themselves. This classification created understanding about the real life socio-economic status of the villagers. It is hoped that this information will provide valuable insights with respect to future planning and rural development projects suiting the real life needs of villagers. This information may also prove helpful to agricultural scientists to develop such cost efficient technologies, which can be adopted by majority of the farmers.

Under this PRA technique, seven local key informants were identified who had in-depth idea about the villagers. They were asked to classify all households of Bar Ka Nagla village into different wealth categories according to their own criteria of classification. The criteria and categories followed by the key informants are given in the table below:

Table 12. Wealth ranking categories, criteria and households

Category	No of Households	% Households	Criteria
Rich farmers	10	20	>5 ha land, livestock, Govt. job, outside business
Medium farmers	12	24	1-2 ha land, livestock, Job
Poor farmers	12	24	< 1 ha land, livestock, temporary private job
Landless labourers	16	32	Livestock, Labour work

Total number of households - 50

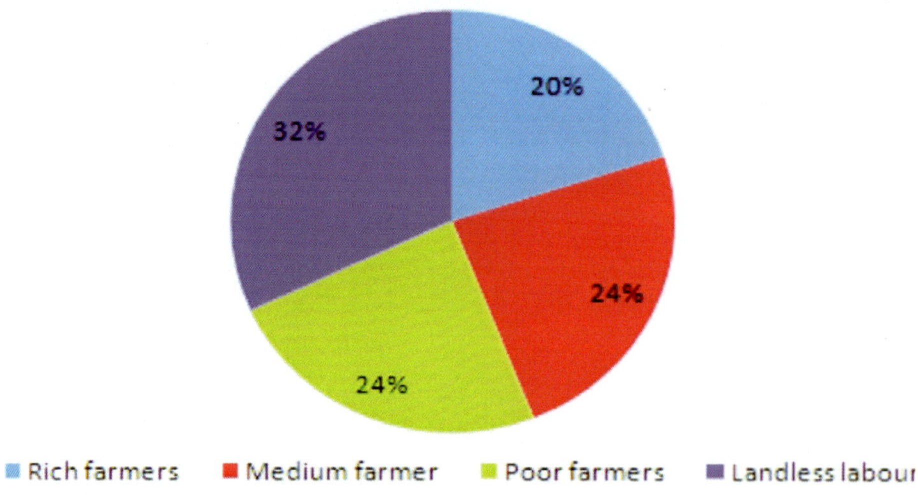

Figure 7. Percentage of households under different wealth status

Out of the 50 households of Bar ka Nagla village, only 10 households came under the rich farmers category (20%), 12 households under the medium farmers category (24%) and 12 households under poor farmers category. Sixteen households, categorized as landless labourers did not have any arable land. They were solely based on their physical labour and small scale livestock farming. Therefore, only 60 per cent of the household of this village were having land and any technology to be developed, should concern about them. This situation highlights the need for development of local agro-based industries which will not only ensure regular employment to the landless rural populace but would also lead to increased crop and enterprise diversification. It was noted that the study village and surrounding areas have good potential for the development of agro-industries. Potato is the main Rabi crop of this region, but value addition and processing are virtually absent. The region has good agro-climatic conditions to grow mushroom and flower crops, such as rose, marigold, chrysanthemum and gladiolus. The twin cities of Agra-Mathura have good demands for mushrooms and flowers. In spite of the presence of CIRG, Makhdoom, commercial goat farming has not been taken up on commercial scale.

5.12 Livelihood Analysis

Using the wealth ranking method, the entire Bar Ka Nagla village households were divided into four categories, *viz.*, rich farmers, medium farmers, poor farmers and landless labourers. The livelihood statuses of each of the four categories were assessed by selecting a representative house from each group. However, the landless labour, constituting 32 per cent of the household. The percentage contribution of different income sources and expenditure pattern of all the four categories are represented below graphically.

Income and expenditure pattern

A) Rich farmer (**Key Informant:** Deewan Singh)

Deewan Singh had 10 ha of land. The gains from crop production in 10 ha of land constituted about 40 per cent (approx 2.5-3 lakhs) of the total income. One male member of his family worked in CIRG, Makhdoom and his salary constituted about 50 per cent of the total family income. Ten per cent of his income accrued from livestock rearing. The family spent approximately 31 per cent income for crop production (Rs. 60,000 each per year). Similarly, household expenditure accounted for 25 per cent of the total expenditure. Healthcare, education of children and livestock rearing constitute 5, 4 and 2 per cent of the total expenditure, respectively. It was surprising to note that crop production made 40 per cent of total income and 31 per cent of total expenditure went to farming. Thus, income from crop production was very low, as compared to that from livestock and job salary. This revealed that agriculture was not remunerative and farmers need to revisit their approach. Greater focus should be

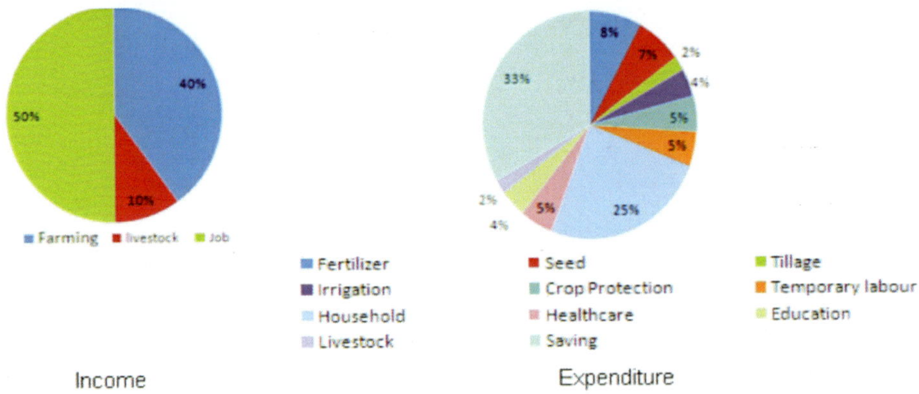

Figure 8. Income and expenditure of rich farmers

laid on mixed farming, *i.e.*, combining crop and animal husbandry for realizing higher profits on a sustainable basis.

B) Medium farmer (**Key Informants:** Vijay Singh)

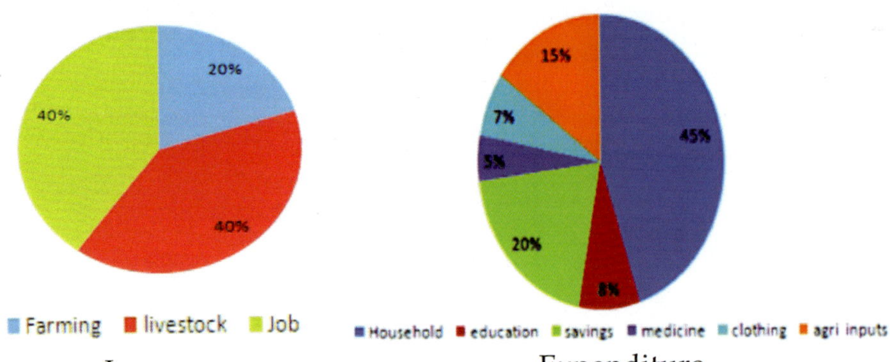

Figure 9. Income and expenditure of medium farmers

Vijay Singh had 2 ha of land and 20 per cent of his income was from crop production and 40 per cent from livestock. Additionally, 40 per cent of his total income came from private job. His elder son was working in a small petha factory, in Agra. Forty five per cent of his expenditure was for household purpose, 15 per cent for agri-inputs like seeds, fertilizer, labour, etc. and 5 per cent for healthcare. He had 2 children who were going to school. That is why his 8 per cent of the expenditure was for educational purpose. Anyhow, he was managed to save 20 per cent.

C) Poor farmer (**Key Informant:** Kailash)

Kailash had 0.5 ha of land and 15 per cent of his total income was coming from each farming and goat rearing. Although Mr. Kailash had only 0.5 ha of arable land, he managed to make 15 per cent of his total earnings from goat farming. This implied that goat farming was a profitable venture for small and marginal farmers, as it required less initial investment and maintenance cost but gave higher returns in subsequent years leading to higher benefit: cost ratio. Besides farming, he also work, as a part-time labour to others field. Therefore, 35 per cent of his total income was coming from labour work. Out of his total expenditure 50 per cent was spent for household, 10 per cent each for agri inputs and education. Amount spent in medicine constituted 10 per cent of total Income Expenditure expenditure. He didn't have any savings.

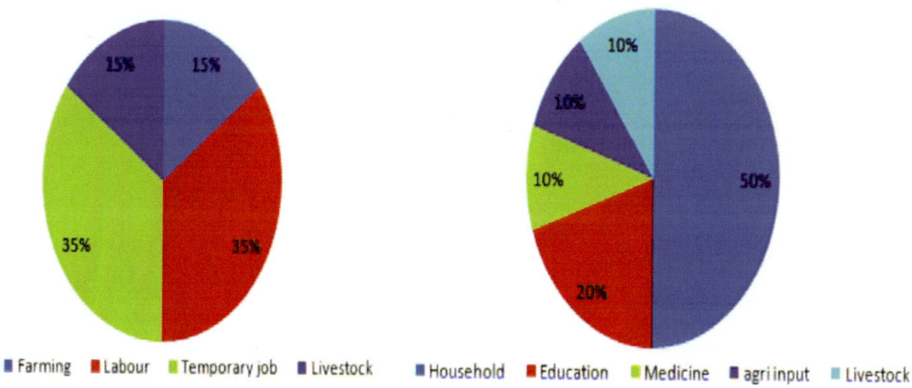

Income Expenditure

Figure 10. Income and expenditure of poor farmers

D) Landless labourers (**Key Informant:** Maniram Singh)

Maniram Singh had no farm land. He was working as labour in sugar mill in Agra. So, he earned 65 per cent of his total income by working as a labour in this mill. He earned 25 per cent from livestock sector. 10 per cent of his income was generated from cultivation on leased land. Out of his total expenditure, he spent 50 per cent for household maintenance, 20 per cent each for livestock and education and 10 per cent for medicine. He saved nothing from his income source. Some more samples representing different landless labourers were analysed. These landless labourers were unable to save anything. Although parents in these families were uneducated but boy children were going to school. Girls were not going to school which might be due to shortage of money and less importance given to girl child in rural societies. This trend reflected the age old discrimination of girl child in traditional Indian society. Anyway, emphasis on children's education implied that rural people have realized the potential of education for socio-economic empowerment in the future.

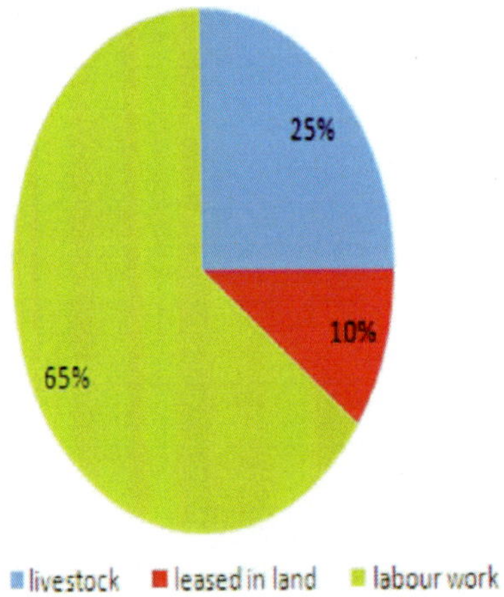

Income

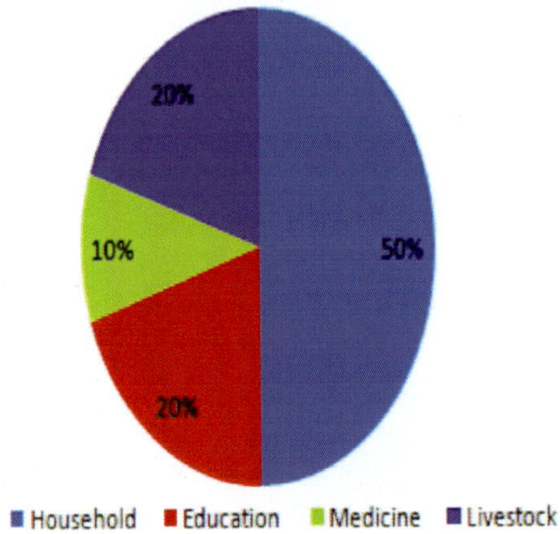

Expenditure

Figure 11. Income and expenditure of landless labourers

Table 13. Livelihood analysis of landless labourers

KI	Income	Expenditure	Family	Education	Livestock
Pooram Singh	Labour work (75%) Livestock (25%)	Household (50%) Livestock (20%) Education (20%) Medicine (10%)	6 (3 adults, 3 children) 2 boy & 1 girl	Adults are uneducated but children are going to school	2 Buffaloes
Maniram Singh	Labour (80%) Livestock (20%)	Household (60%) Livestock (15%) Education (15%) Medicine (10%)	8 (4 adults, 4 children) 2 boy & 2 girl	Boys are going to school but girls are uneducated	1Buffales, 4 goats

Suneheri	Leased in land (35%) Labour (45%) Livestock (20%)	Household (55%) Education (20%) Medicine (10%) Livestock (15%)	6 (4 adults, 2 young son)	One son going to college & another one work in other's land.	2 pigs

Houschold composition and migration pattern

The family tree of Pooran Singh clearly showed that out of his two sons, one had migrated to Delhi, as he is serving in Indian army. However, he is not a permanent migrant and maintains close ties with his family in the village. He also sends remittances to the family and in this way improves the income generation of his family. Another son of Mr. Singh is working in CIRG, Makhdoom and contributes his salary to the family. In general, most of the youth in the village are interested in having government jobs for ensuring their livelihood. Most of them do not see a right future in farming and livestock rearing. This is an alarming situation and reminds us about our ailing agricultural sector. Besides Mr. Singh's family, there was no other instance of outward or inward migration.

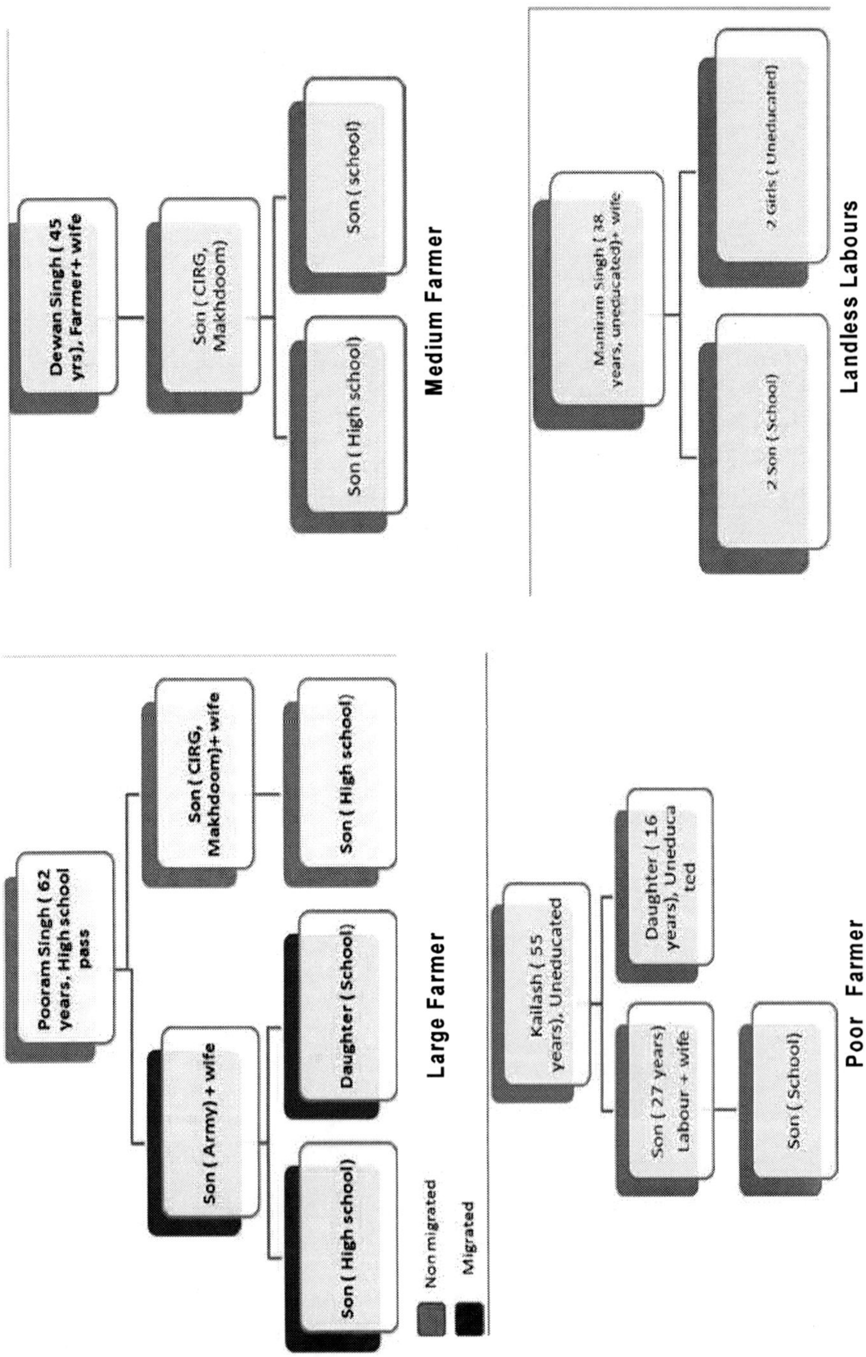

Figure 12. Household composition and migration pattern of different categories

5.13 Seasonal Calendar and Analysis

Key informants : Vijay Singh, Kailash, Sitaram, Umabati and Tikam Singh

In rural areas, all the activity of the people completely depends upon the season. Different seasons bring different types of problems and opportunities, and the villagers have their own methods of coping with them. A seasonal calendar depicts the seasonal variation with respect to a number of variables, such as production of agricultural crops, disease in crops and livestock, market prices, consumer demand, fodder availability, pest attack in crops, labour strategies, fuels wood availability, etc. It is also called as seasonal diagram.

The seasonal calendar focuses on local livelihood system. They saw month by month variation of rainfall, cropping patterns, water use, livestock fodder, income, debt, migration, health status of livestock, diseases and so on. The main activities, problems and opportunities of Bar ka Nagla village were identified by using seasonal calendar. It depicts time-to-time crop related operations being carried out in the existing farm situation. The common main crops in the village are potato, wheat and mustard (Rabi) and bajra (Khariff). .Seasonal analysis helps in identifying the period which are critical with respect to labour demand, pest and disease problems and availability of fodder.

Plate 9. Seasonal activities of Bar ka Nagla village

Similarly, analysis of seasonal calander in livestock helps to give an overview of scenario of livestock activities and health status in overall. The most common disease of livestock in the village was FMD in cattle and buffalo, and diarrohea as well as parasite infestation in young kids of goat. Farmers are well aware of improved crop varieties but they are slow in adopting latest technologies. Labour availability is inadequate throughout the year. The most remunerable activities in the village are selling of potato, wheat and mustard.

Table 14 Seasonal calendar of major crops and their practices in Bar ka nagla village

CROPS	NOV	DEC	JAN	FEB	MAR	APR	MAY	JUN	JUL	AUG	SEP	OCT
Potato	E, 1-IR, PS	2-IR, WD	3-IR,F	HR								
Wheat	SW, F	F, 2- IR, 3- IR	4-IR	5-IR		HR						PSW
'Mustard	1-IR, F			HR								
Bajra + Sesamum						FPR		IR	HR			

(*Note*: E- Earthing up, 1 IR- First Irrigation, 2-IR-Second irrigation, 3-IR- third Irrigation, IRIV- fourth irrigation, HR-Harvesting, WD- Weeding, SW- Sowing, PSW- Post sowing, FPR-Field Preparation, F- fertilizer application, PS-pest spray)

Table 15. Seasonal calendar of major crops and their problems in Bar ka Nagla village

CROPS	NOV	DEC	JAN	FEB	MAR	APR	MAY	JUN	JUL	AUG	SEP	OCT
Potato		Frost Injury Late blight Potato Leaf roll virus										
Wheat	Termite Problem			Karnal Bunt (Kala dabba)								
Mustard			Aphid (chepa)									
Bajra + Seasamum										Smut (Khandwa)		

Table 16. Seasonal calendar of livestock health management practices in Bar ka Nagla village

Livestock	Nov	Dec	Jan	Feb	Mar	Apr	May	Jun	Jul	Aug	Sep	Oct
Cow							Breeding season (NS)					
				Calving Season								
Buffalo					Breeding season (NS)							
		Calving season										
Goat									De-worming			
Sheep									De-worming			

(*Note:* NS: Natural Service)

Table 17. Seasonal calendar for livestock problems

Livestock	Nov	Dec	Jan	Feb	Mar	Apr	May	Jun	Jul	Aug	Sep	Oct
Cow	FMD							Parasitic infestation, Mastitis				
Buffalo	FMD							Mastitis	Bloat			
Goat								Dysentry, Bloat, Worm infection				
Sheep								Parasitic infestation				

(*Note:* FMD-Foot and Mouth Disease)

5.14 Gender Disaggregated Activities

Key informants : Somvati Devi, Saroja bibi, Umabati Devi, Rajan Singh, Kisan Singh

According to information revealed by the key informants, women in Bar Ka Nagla contributed mainly to livestock management and home management. Ladies would feed the livestock, milk the animals in the morning, take the animals to the fallow land, bath them at noon, milk them again in the evening and cut fodder for the animals. Landless women and women whose husbands had little land, also worked as agricultural labourers either within the village or in nearby towns. Men are mainly involved in field activities as well as selling milk.

Plate 10. Gender Disaggregated Activities

Table 18. Gender disaggregation seasonal calendar of Bar ka Nagla village

CROPS	NOV	DEC	JAN	FEB	MAR	APR	MAY	JUN	JUL	AUG	SEP	OCT
Potato	E, 1-IR, PS(M)	2-IR, WD(M)	3-IR, F (M)	HR (F+C)								
Wheat	SW, F(M)	F, 2- IR, 3- IR (M)	4-IR (M)	5-IR (M)		HR (F)						PSW (M)
Mustard	1-IR, F(M)			HR (F)								
Bajra + Sesamum						FPR(M)		IR(M)	HR(F)			

(*Note:* M-Male,F-Female, C-Children, E- Earthing up, 1 IR- First Irrigation, 2-IR-Second irrigation, 3-IR- Third Irrigation, IRIV-Fourth irrigation, HR-Harvesting, WD- Weeding, SW- Sowing, PSW- Post sowing, FPR-Field Preparation, F- Fertilizer application, PS-Pest spray)

Table 19. Gender disaggregated Seasonal calendar (Livestock)

Livestock	Nov	Dec	Jan	Feb	Mar	Apr	May	Jun	Jul	Aug	Sep	Oct
Milking							Women					
Fodder harvesting							Men/Women					
Grazing							Women					
Disease management							Men/Women					
Marketing							Men					
Cleaning							Women					

5.15 Time Line

Key Informants : Ramprakash, Vikram Singh, Deewan Singh and Jaswant Singh

Time line is a tool of PRA technique, used to know the history of major remembered events in community and their significance. It indicates the causal link between past and present. The purpose of this tool is to obtain historical account of changes in demography, socio-economic condition, communication, social relationship and interaction, technology diffusion and adoption, etc. Information regarding the important events, like the development in agriculture, animal husbandry are collected from the key informants, preferably elderly people of the village and presented in the table.

Table 20. Time line of major events in Bar Ka Nagla village

S.No.	Agricultural Events	Year
1.	Canal irrigation	1961
2.	Chipka-Rohini varieties of Mustard	1981
3.	Tube well	1981
4.	Potato plucker	1981
5.	Tractor (Massi, Saroj,Iceor)	1986
6.	Tomato cultivation	1986
7.	Urea fertilizer	1986
8.	Sowing machine	1999
9.	Hand pump	2005
10.	Motor	2006
11.	Harrowing machine	2006
12.	Harvesting machine	2006
13.	Disc harrow	2007
14.	Seed drill	2007
15.	Land laveller	2008
16.	Yoke	2008
17.	Patela	2009
18.	3797 variety of potato	2009
19.	Thrasher	2009
20.	Mobile	2009
21.	Computer	2010
22.	Motor cycle	2010
23.	Seed cum fertilizer drill	2011

5.16 Time Trend

Key informants : Jaswant singh, Yadarn, Madan Lal, Ujjam, Deewan Singh

Time trend is a simple PRA technique, usually depicted in the form of graph (bar/line) to show the trend of crop/animal production, commodity prices, human/cattle population, etc. The specific objectives of this tool are to identify the changes/fluctuations that have occurred over a period of time in the variables influencing village life. For the purpose of the present PRA study, the data were collected for crop productivity of agricultural crops (Rabi viz. Mustard, Potato, Wheat and Khariff viz. Bajra) and milk production from the buffaloes over the last 5 years and are presented graphically below.

a. Agricultural crops

The trend analysis of all the crops productivity shows a decreasing trend from 2005-06 to 2010-11. This is due to mostly saline hard water problem in the village and lack of awareness regarding the control measures of saline soil.

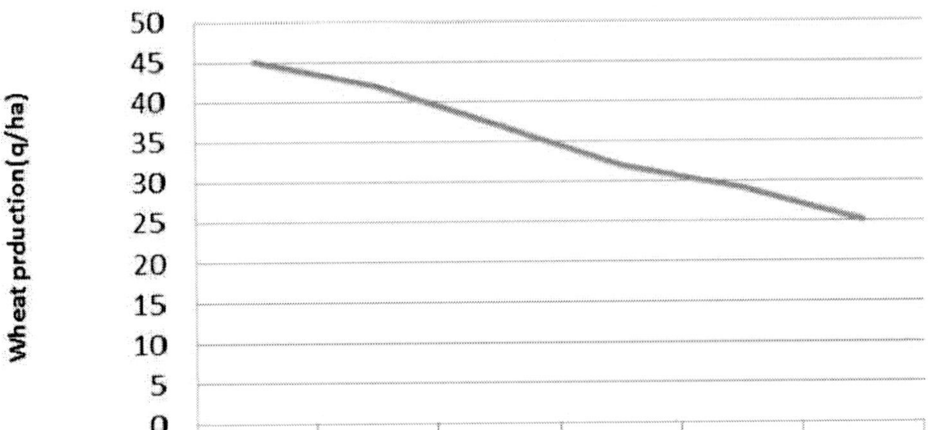

	2005-06	2006-07	2007-08	2008-09	2009-10	2010-11
Series1	45	42	37	32	29	25

Figure 13. Time trend analysis for Wheat

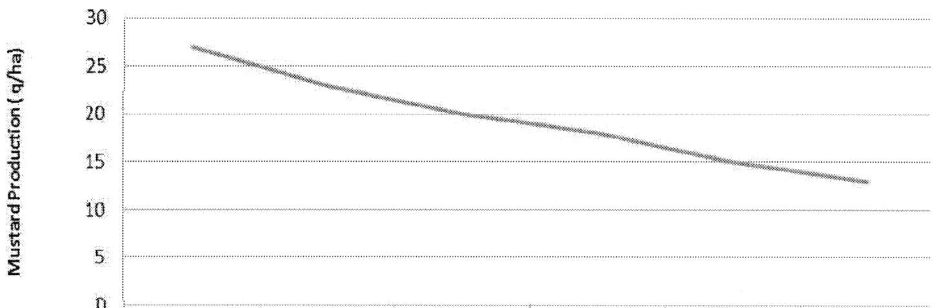

	2005-06	2006-07	2007-08	2008-09	2009-10	2010-11
Series1	27	23	20	18	15	13

Figure 14. Time trend analysis for Mustard

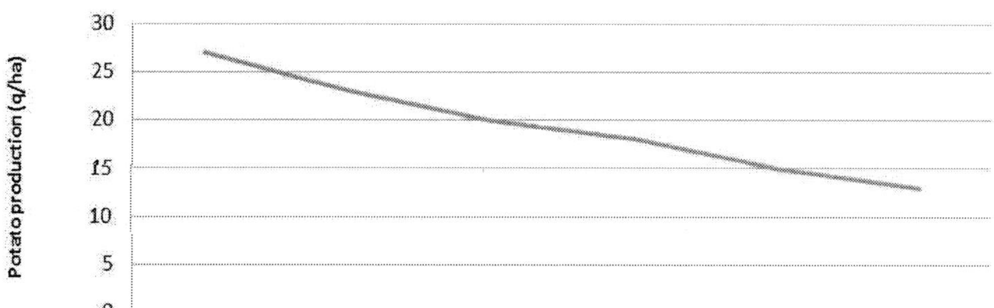

	2005-06	2006-07	2007-08	2008-09	2009-10	2010-11
Series1	27	23	20	18	15	13

Figure 15. Time trend analysis for Potatoes

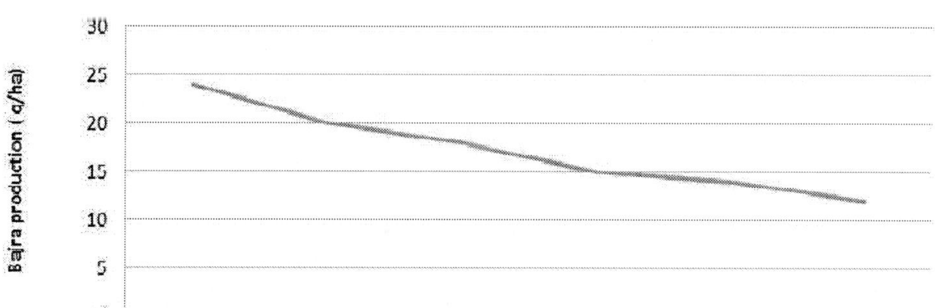

	2005-06	2006-07	2007-08	2008-09	2009-10	2010-11
Series1	24	20	18	15	14	12

Figure 16. Time trend analysis for Bajra

b. Milk productivity

Again the trend analysis of the buffalo milk productivity shows an decreasing trend from 2005-06 to 2010-11. This was due to the reason that animals were drinking the saline/hard water and also suffer from FMD, like diseases due to improper health care, like vaccination.

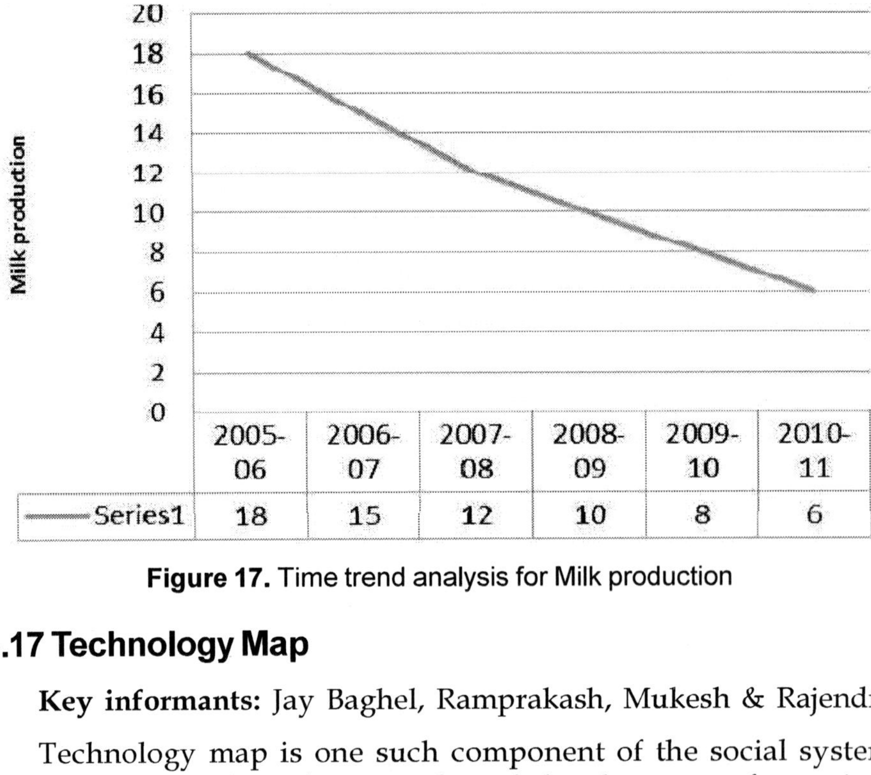

Figure 17. Time trend analysis for Milk production

5.17 Technology Map

Key informants: Jay Baghel, Ramprakash, Mukesh & Rajendra

Technology map is one such component of the social system, which forms the base for growth and development of a society. Technology is continuously changing and the social has to adopt the technological changes to keep pace with the speed of development and those individual/ society which don't adopt successful technology becomes laggards and failed to receive the benefits of this technologies. The research system is engaged in continuous development of newer technologies for agriculture, animal husbandry, fisheries, horticulture and other enterprises to enhance the productivity but the adoption of these technologies at field level holds the key for raising production and productivity. The technology adoption behaviours with respect to various aspect of rural livelihood can portray the strength and lacunae in the villages and can guide the researcher in developing appropriate action plant for introducing new technologies in the village.

Technology map provides an in-depth analysis of a technology adapted by the farmers. It works as a comprehensive guide for the other farmers also, who would like to adopt in course of time. Technology map depict the technology adaption behaviour of the villagers. It urge the technology that have been adapted, discontinued, over adapted, reinvented or rejected by the farmers of a village.

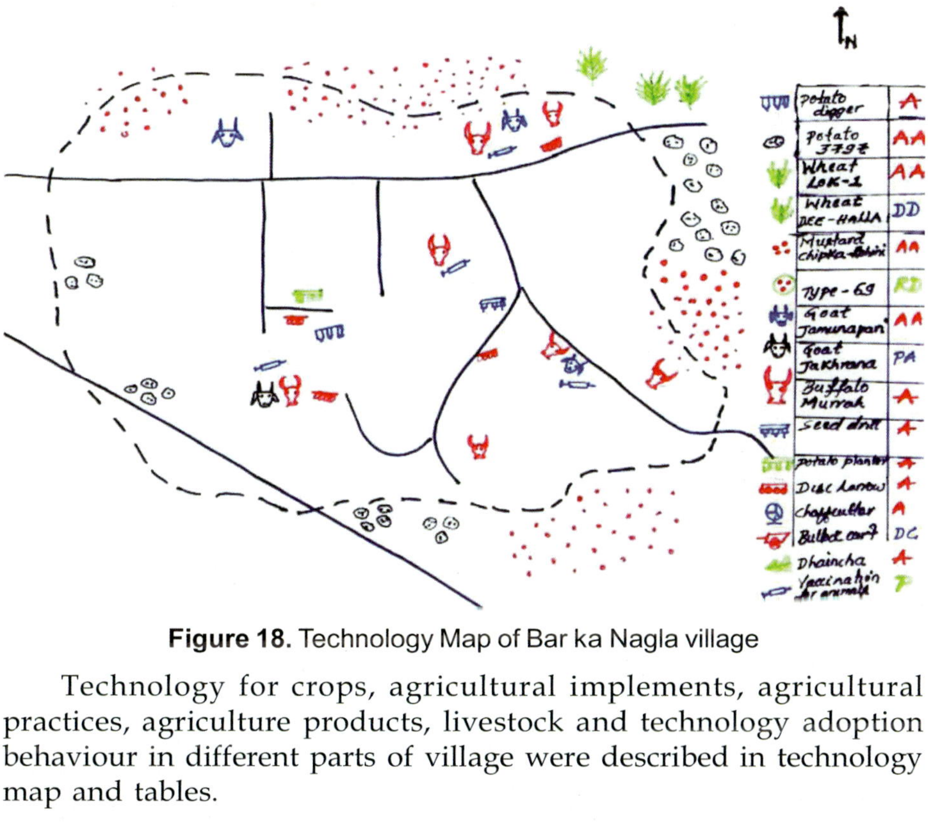

Figure 18. Technology Map of Bar ka Nagla village

Technology for crops, agricultural implements, agricultural practices, agriculture products, livestock and technology adoption behaviour in different parts of village were described in technology map and tables.

The main occupation of villagers is agriculture. The farmers were growing Bajra + Sesamum (intercropping) during khariff and three major crop in rabi viz; Potato, Wheat and Mustard. The variety and their adopted behaviour by farmers for different crops were presented in tables. Recently, they had adopted an alternate way of livelihood i.e. Rearing of Goat and Buffalo for milk and assured income. The majority of goat breed were Jamunapari since this breed has wider adaptability and high milk yield over other reared breeds, like Jhakhrana and Barbari.

Table 21. Technology table for potato, wheat and mustard crops

S.No.	Crop	Technology	Status	Reason
1.	**Potato** 	3797	AA	Good tuber colour as well as cooking quality, heat tolerance, highest price in market, tolerance to rotting.
		Kufri Sutlej	A	Good cooking quality, heat tolerance, good price in market, tolerance to rotting.
		Spl-1	A	Frost tolerance, high yield
		Local (Self produce tuber)		DD Poor yield, susceptible to blight and aphids, problem of rotting.
2.	**Wheat** 	Lok-1	AA	Good cooking quality, bunt tolerance, early maturity, high market value.
		2009	A	Higher yield, low marketing price.
		HD-2329	DC	Late maturity
		Dee Halah	DD	Low marketing price, poor yield, high burnt infection.
3.	**Mustard** 	Chipka Rohini (Farmer Selection)	AA	High seed and oil yield, early maturity, aphid, white rust tolerance, resistance to shattering.
		Type-69	RD	Low yield, shattering problem

(Where, A: Adopted; DD: Disenchantment discontinuance, AA: Active adoption, RD: Replacement discontinuance, DC: Discontinuance, AA: active adoption, PA: Passive adoption).

Plate 11. Triangulation with farmers for Technology Map

Table 22. Technology table for farm machineries & implements, agricultural practices & products, and livestock

S.No.	Technology	Status	Reason
Farm machineries and implements			
1.	Cultivator	A	Rapid and single step, easy field preparation than traditional method.
2.	Disc harrow	A	To get fine tilth in soil (for wheat)
3.	Potato planter	A	Requires less labour requirement, timely sowing.
4.	Potato digger	A	Less labour
5.	Knapsack sprayer	A	Uniform application of plant protection material
6.	Seed drill-cum - fertilizer drill	A	Ease and convenience in sowing seeds and fertilizer application
7.	Tractor	A	For agricultural practices and high capacity transportation.
8.	Land leveller - tractor drawn	A	For rapid levelling, better uniformity.
9.	Chaff cutter	A	For cutting feed in to small parts.
10.	Bullock cart	DC	No bullocks, low capacity transportation.
11.	Diesel engine	A	Pumping out irrigation water, electricity problem.
Agricultural practices and products			
1	Mechanization in potato cultivation	AA	Less labour requirement

2	Application of FYM	A	Management of soil health, efficient utilization of farm by product, reducing chemical fertilizer requirement
3	Frost management in potato	A	Reduce the risk of frost injury
4	Drip irrigation	NA	Lack of awareness
5	Soil testing	NA	Lack of awareness
6	Cow dunk cake	NA	For fuel purpose
7	Ag. Magazines	NA	Lack of awareness
Livestock			
1.	Donkey and Mule	A	Transportation
2.	Goat - Jamunapari	AA	High milk yield, shorter lactation length, easily adopted in this region.
3.	Goat - Jakhrana	PA	Require more feeds.
4.	Goat-Barbari	A	Newly adopted
5.	Murrah buffalo	A	Income from milk and for transportation.
6.	Artificial insemination	NA	Lack of awareness
7.	Hay making	A	For sustainable feeding
8.	Poultry farming	DC	Religious taboo.
9.	Apiculture	NA	Lack of awareness
10.	Vaccination for animals	PA	Lack of awareness

Where, A - Adopted; DD - Disenchantment discontinuance, AA - Active adoption, RD - Replacement discontinuance, DC – Discontinuance, AA - Active adoption, PA - Passive adoption, NA - Not adopted

5.18 Matrix Ranking

Key informants : Jay Baghel, Ramprakash, Mukesh & Rajendra

Matrix ranking is use to learn from the people what they perceive about particular technology i.e. negative and positive feelings in both absolute and relative terms. This helps to develop a better focus on the transfer of technology programme for the welfare of poor segment of society. Ranking helps in knowing about the preference of individuals, regarding a number of aspects of life. It can tell more about what the individual prefers most and least. Matrix ranking helps in identifying a number of criteria related to some particular items and in ranking this items on the basis of identified criteria.

Matrix ranking was applied in the village Bar ka Nagla for getting farmers' relative preference for different varieties of potato, wheat, mustard and also, different goat breeds. From the results of the matrix ranking for different varieties of potato, it was found that variety 3797 was the highest preferred one, when compared to the remaining two adopted varieties, viz., Sutlej and special-1. It was clearly understood that the farmers had been mainly concerned with colour, taste and heat tolerance which influence the high market price. Among the three adopted varieties of wheat viz; Loc-1, 2009 and HD-2329, the second two were almost equally preferred by the farmers but the first one has advantage on cooking quality, burnt tolerance, early maturity and high market price. Two mustard varieties were grown by farmers viz; Chipka-Rohini and Type-69. Out of these two Chipka-Rohini was found to be the best one in terms of quantitative, qualitative and economic aspects. Besides the crops, the villagers have also adopted the goat and buffalo for alternate source of livelihood. We studied the matrix ranking for goat only because there were two distinct breed and all buffaloes were murrah breed. Among the two breed of goat Jamunapari is superior, farmers were concern with both milk and meat from goat.

Table 23. Matrix ranking of potato varieties

Indicator	KI	Potato varieties					
		3797		Sutlej		Special-1	
		Rank	Score	Rank	Score	Rank	Score
1. Colour	KI-1	A	3	B	2	C	1
	KI-2	B	2	A	3	C	1
	KI-3	A	3	B	2	C	1
Total score			**8**		**7**		**3**
2. Tuber Yield	KI-1	C	1	B	2	A	3
	KI-2	B	2	C	1	A	3
	KI-3	C	1	B	2	A	3
Total score			**4**		**5**		**9**
3. Cooking	KI-1	A	3	B	2	C	1
quality (Taste)	KI-2	A	3	B	2	C	1

	KI-3	A	3	B	2	C	1
Total score			**9**		**6**		**3**
4. Frost	KI-1	C	1	B	2	A	3
Tolerance	KI-2	C	1	A	3	B	2
	KI-3	A	3	C	1	B	2
Total score			**5**		**6**		**7**
5. Heat	KI-1	A	3	B	2	C	1
Tolerance	KI-2	A	3	B	2	C	1
	KI-3	B	2	A	3	C	1
Total score			**8**		**7**		**3**
6. Market Price	KI-1	A	3	B	2	C	1
	KI-2	A	3	B	2	C	1
	KI-3	B	2	C	1	A	3
Total score			**8**		**5**		**5**
Overall score and final rank			**42**		**36**		**30**

Table 24. Matrix ranking of wheat varieties

Indicator	KI	Wheat varieties					
		Loc-1		2009		HD-2329	
		Rank	Score	Rank	Score	Rank	Score
1. Cooking quality	KI-1	A	3	C	1	B	2
	KI-2	B	2	B	2	C	1
	KI-3	A	3	A	3	C	1
Total score			**8**		**6**		**4**
2. Yield	KI-1	C	1	A	3	B	2
	KI-2	C	1	B	2	A	3
	KI-3	C	1	A	3	B	2
Total score			**3**		**8**		**7**
3. Bunt tolerance	KI-1	A	3	C	1	B	2
	KI-2	A	3	B	2	C	1
	KI-3	A	3	C	1	B	2
Total score			**9**		**4**		**5**
4. Early Maturity	KI-1	A	3	B	2	C	1
	KI-2	A	3	B	2	C	1
	KI-3	A	3	B	2	C	1
Total score			**9**		**6**		**3**

5. High market price	KI-1	A	3	C	1	B	2
	KI-2	B	2	C	1	A	3
	KI-3	A	3	C	1	B	2
Total score			**8**		**3**		**7**
Overall score and final rank			**37**		**27**		**26**

Plate 12. Matrix ranking in Bar ka Nagla

Table 25. Matrix ranking of Mustard varieties

Indicator	**KI**	**Mustard varieties**			
		Chipka Rohini		**Type-69**	
		Rank	**Score**	**Rank**	**Score**
1. Seed Yield	KI-1	A	2	B	1
	KI-2	A	2	B	1
	KI-3	A	2	B	1
Total score			**6**		**3**
2. Oil percentage	KI-1	A	2	B	1
	KI-2	B	1	A	2
	KI-3	A	2	B	1
Total score			**5**		**4**
3. Shattering tolerance	KI-1	A	2	B	1
	KI-2	A	2	B	1
	KI-3	A	2	B	1
Total score			**6**		**3**
4. Early Maturity	KI-1	A	2	B	1
	KI-2	A	2	B	1

	KI-3	A	2	B	1
Total score			**6**		**3**
5. Aphid Tolerance	KI-1	A	2	B	1
	KI-2	B	1	A	2
	KI-3	A	2	B	1
Total score			**5**		**4**
Overall score and final rank			**28**		**17**

Plate 13. Matrix ranking of Mustard varieties

Table 26. Matrix ranking of goat breeds

Indicator	**KI**	**Goat breeds**			
		Jamunapari		**Jhakhrana**	
		Rank	**Score**	**Rank**	**Score**
1. Climatic adaptability	KI-1	A	2	B	1
	KI-2	A	2	B	1
	KI-3	A	2	B	1
Total score			**6**		**3**
2. Milk yield	KI-1	A	2	B	1
	KI-2	A	2	B	1
	KI-3	A	2	B	1
Total score			**6**		**3**
3. Meat yield	KI-1	B	1	A	2

	KI-2	B	1	A	2
	KI-3	B	1	A	2
Total score			**3**		**6**
4. Maintenance cost	KI-1	A	2	B	1
	KI-2	B	1	A	2
	KI-3	B	1	A	2
Total score			**4**		**5**
Overall score and final rank			**19**		**17**

5.19 Consequence Diagram

Key informants : Om Prakash Sharma, Nawab Singh Baghel and Narayan Singh

Consequence diagram, also known as impact diagram is a PRA technique that depicts the positive and negative impacts of an adapted technology in a certain region. Consequence diagram has been presented below for the goat rearing, potato cultivation and adaptation of mustard variety.

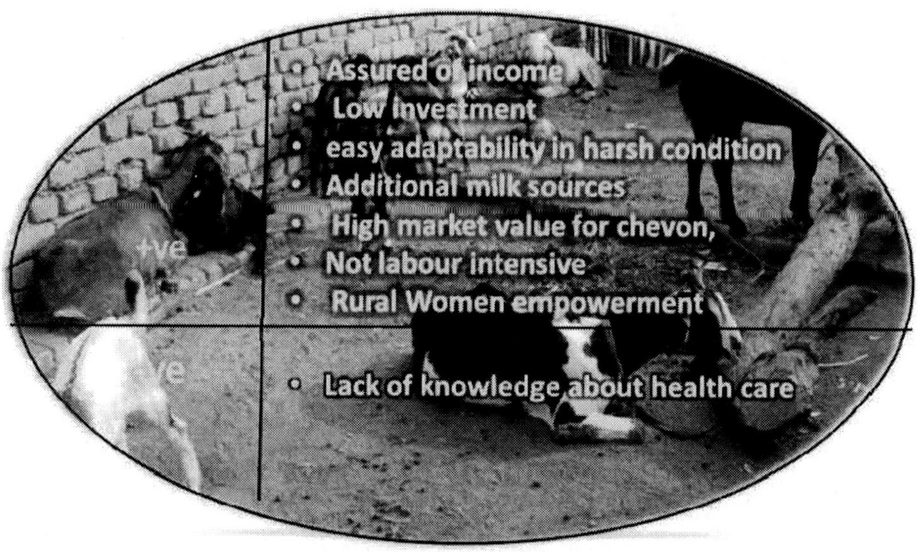

Figure 19. Consequence diagram for goat rearing

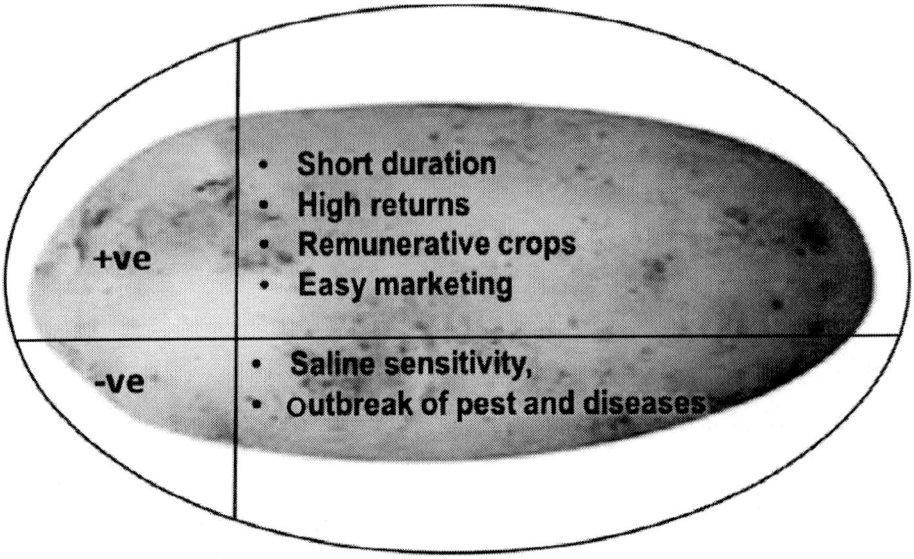

Figure 20. Consequence diagram for potato cultivation

Figure 21. Consequence diagram for Mustard variety - CR

5.20 Problem Identification Techniques

Key informants : Seeta Ram, Pooran Singh, Paramjit Singh, Vijay Singh, Hari Das, Om Prakash Sharma, Nawab Singh Baghel and Narayan Singh

After collecting all the information about the technologies and resources available and adopted in Bar ka Nagla village, we identified the problems that were being faced by the farmers, especially pertaining to the agricultural and animal husbandry sector. A sample of 30 farmers was identified, using the snowball technique to determine the problems. A total of eight important problems were identified in the village and the farmers were then asked to rank them according to their perceptions about the severity of the problem. Maximum of eight ranks were given by the farmers. Some of the farmers ranked all the problems, while some ranked only for problems relating to agriculture or livestock. Frequency of each rank was calculated. Farmers were also enquired about the average loss (per cent damage) for the respective problems. Average monetary loss for each problem was also calculated. Then, the Rank Based Quotient (RBQ) and Value Based Index (VBI) were calculated for each problem.

Rank Based Quotient (RBQ)

$$RBQ = \sum fi(n+1-i)x100/Nxn$$

Where, i = Concerned ranks (1 to 8)

n = Number of ranks (up to 8)

N = Total number of farmers contacted for problem identification (N=30)

f_i = Frequency of ith rank (Number of farmers reporting particular problem under i^{th} rank)

Table 27. Problems identified by the villagers of Bar ka Nagla

S.No.	Problem	Rank								RBQ
		1	2	3	4	5	6	7	8	
1.	Saline water (*khaara pani*)	12	8	4	1	5	-	-	-	**40**
2.	High mortality in goats due to diarrhoea	3	7	6	4	3	7	-	-	**20.41**
3.	Low yield in mustard due to aphids (*cheppa*)	2	5	9	4	5	-	5	-	**22.5**
4.	Low yield in potato due to frost injury	1	6	3	9	4	-	7	-	**18.75**
5.	Low yield in wheat due to termite attack	2	5	4	4	9	-	-	6	**15**
6.	Foot and Mouth Disease (FMD) in cattle	-	9	3	7	6	5	-	-	**6.25**

7.	Low milk production in milching animals	1	4	9	6	-	7	3	-	**2.5**
8.	Low yield in bajra due to smut (*kandua*)	-	-	7	7	-	6	6	4	**1.6**

Value Based Index (VBI)

VBI = RBQ x Average loss (in %) due to the particular problem x Value of the main product or bi-product x Area of crops or number of animals.

The problems were then ranked on the basis of VBI (in ascending order), as given in the following table:

Table 28. VBI for the problems identified in Bar ka Nagla

S.No.	Problem	RBQ	Avg. loss per annum (%)	Standardized price (Rs.)	Area under crop/no. of animals	VBI	Final rank
1.	Saline water (khaara pani)	40	–	–	–	–	–
2.	High mortality in goats due to diarrhea	20.41	60	2,000/goat	130	318396000	I
3.	Low yield in mustard due to aphids (chepa)	22.5	15	2,800/qt	20 ha	18900000	II
4.	Low yield in potato due to frost injury	18.75	25	200/qt	15 ha	1406250	IV
5.	Low yield in wheat due to termite attack	15	3	1,000/qt	9 ha	405000	VI
6.	Foot and Mouth Disease (FMD) in cattle	6.25	10	1,500/animal	125	11718750	III
7.	Low milk production in milching animals	2.5	15	25/lt	255	239062.5	VII
8.	Low yield in bajra due to smut (kandua)	1.6	20	650/qt	30 ha	624000	V

5.21 Perspectives of Problems

The problems identified were categorized as problems, arising out of gaps in research in various agricultural disciplines (as discussed above) and as problems arising out of gaps in extension and developmental activities. The main problems related to extension and developmental activities identified are enumerated as follows:

Extension Gap

- Awareness about micro nutrient deficiency
- Awareness about proper sanitation of field and animals
- Awareness about vermicomposting
- Awareness about pest v/s bio-control agents
- Awareness about use of different group of insecticides against sucking pests

Developmental Gap

- Tie-up with the co-operative societies involved in handicrafts and other value addition activities
- Non-availability of fertilizers on time
- Improper implementation of government schemes
- Hoarding of fertilizers leading to a rise in prices

5.22 Problem and Solution tree

Under this PRA tool, all the problems were discussed with the farmers for all the possible reasons. It is short diagrammatic analysis of the problem by the farmers. The possible solution for each problem was also discussed with farmers. These possible solutions were further discussed with the subject matter specialist before the finalization of the tree. The problem with the highest VBI is the top-most researchable problem. In our study village, it is the high mortality level amongst goats due to diarrhoea. The problem-solution trees for the topmost two important researchable problems are given below:

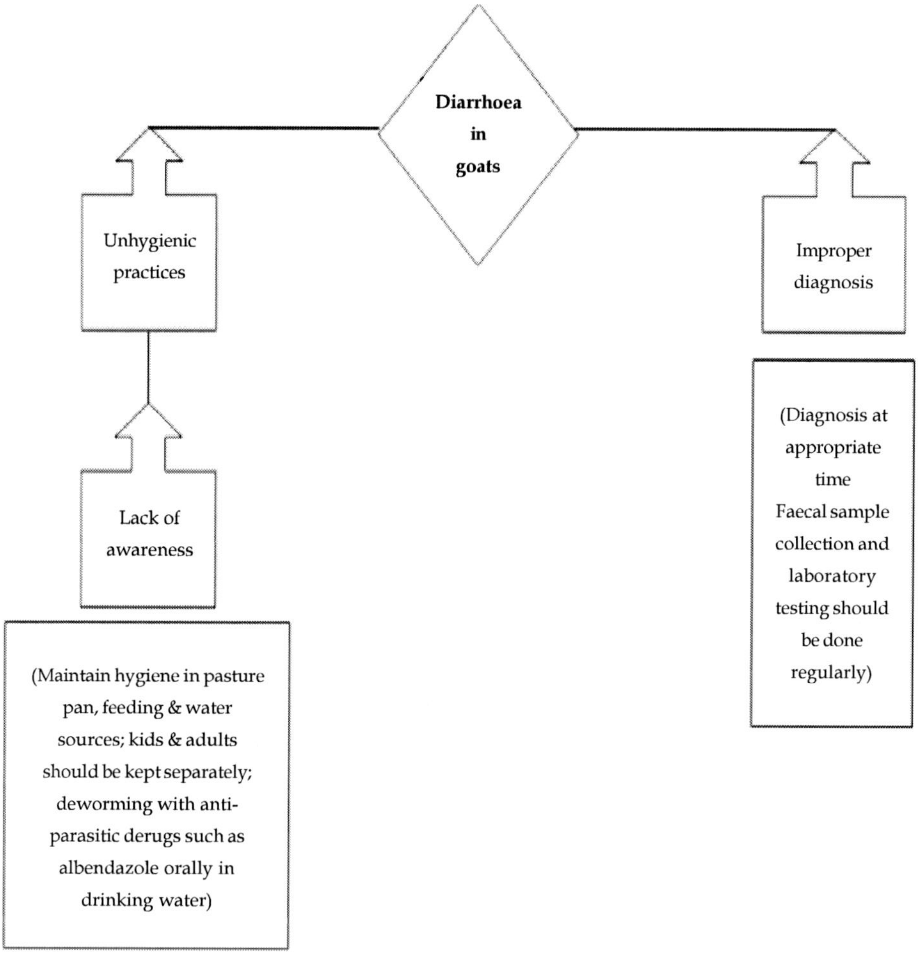

Figure 22. Problem and solution tree for Diarrhoea in goats

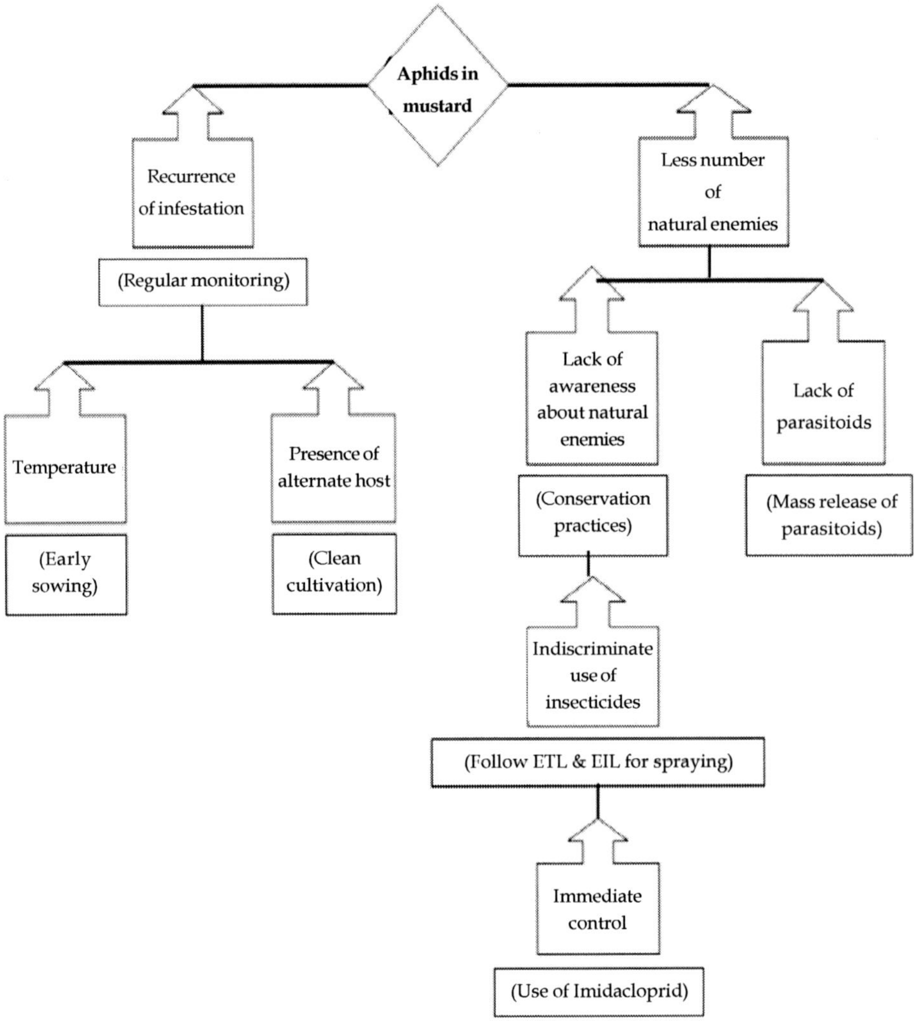

Figure 23. Problem and solution tree for aphids in mustard

Water salinity was observed to be an important problem in our study village. The VBI for the same has not been computed due to the non-availability of average loss (in percentage terms), total area and the associated price. A problem-solution tree for the same has been explained below seeing the criticality of the problem in our village.

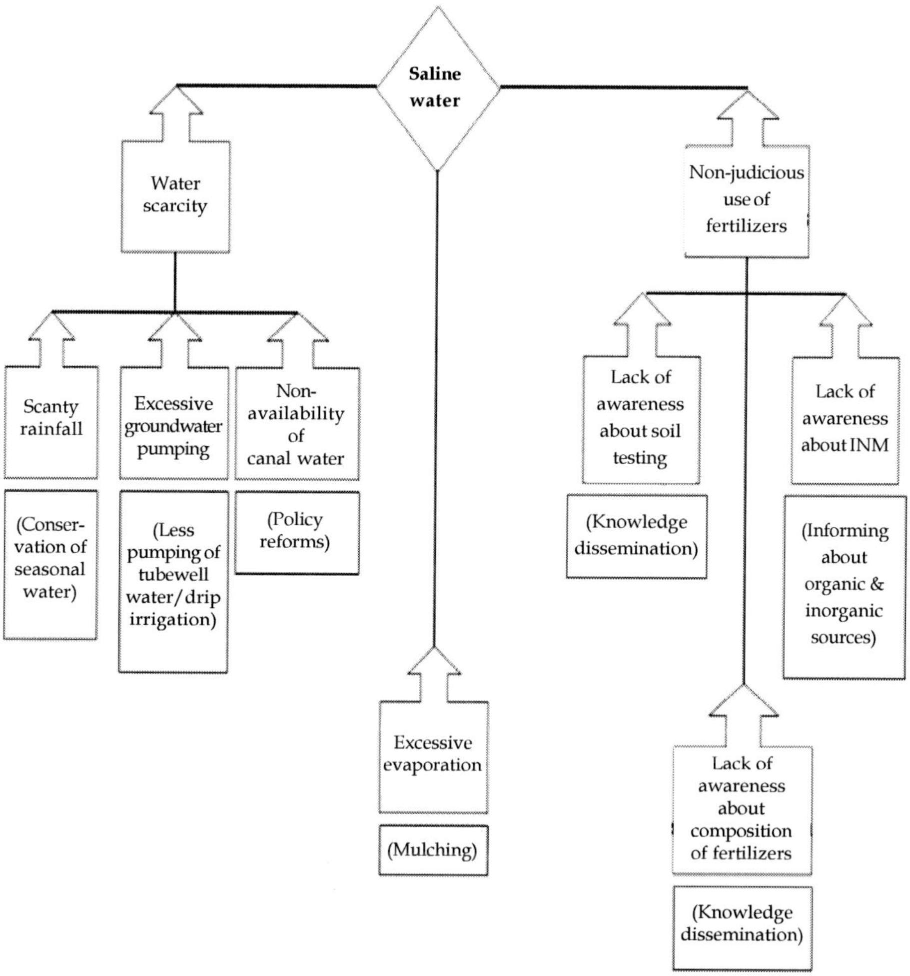

Figure 24. Problem and solution tree for saline water

Chapter 6

Action Plan

Based on the problems and gaps identified, the following action plan, consisting of some of the research proposals is proposed in the table below:

Table 29: Action plan of proposed research proposals

What	Who	When	Where	How	Cost (Rs. lakhs)	Outcome
Studies and survey on the causes of diarrhoea in kids	Veterinary Epidemiologist, microbiologist & parasitologist	2012-14	CIRG, Mathura & IVRI, Bareilly	Laboratory diagnosis	25	Appropriate diagnosis against diarrhoea in kids
Studies on population dynamics of aphid and development of weather-based forecasting model for mustard aphid management	Entomologist & Meteorologist	2012-15	DRMR, Bharatpur	Understanding the bionomics and population dynamics of the pest, in relation to changing weather patterns	30	Effective pest management
Identification of effective chemicals and botanical pesticides	Entomologist	2012-14	IARI, New Delhi	Through pesticide trials	20	Effective pesticides
Development of thermos-	Veterinary microbiologist,	2012-14	IVRI, Bareilly & PD-FMD,	Vector-based vaccine	35	Control of FMD in livestock

table FMD vaccines	Animal Bio-technologist		Mukteswar	development		
Development of IPM practices	Entomologist & pathologist	2012-14	ICRISAT, Hyderabad	Identification of resistant varieties, pesticides and bio-control agents	25	IPM module
Screening of potato germplasm for salt tolerance	Plant breeder	2012-16	CPRI, Shimla	Collection and screening of germplasm under saline soil	40-50	Salt-tolerant plant genotypes

❑❑❑

Chapter 7

Village Seminar

Village seminar "Kisan-Vaigyanik Gramin Sangosthi" was conducted on 21st November, 2011 at Bar ka Nagla village at a convenient place, almost at the centre of the village. Around 57 farmers including women, attended the seminar. Scientists of different divisions of CIRG, Makhdoom attended the village seminar along with FET Programme Co-ordinators, Veterinary officer and technicians; Assistant Manager of Syndicate Bank, Sarpanch and village panchayat members were also present during the seminar. The response to village seminar was very positive since many villagers took active participation in the discussion.

Plate 14. "Kisan-Vaigyanik Gramin Sangosthi" at Bar ka Nagla

A health Camp for livestock was also organized by the FET group with the help of CIRG, Makhdoom. Around 32 villagers were arrived with their cattle and goats for treatment against various health problems, like parasitic infestation, worm infection, bloat, indigestion, etc. We have almost treated more than 40 animals and distributed medicine.

Proper advice for health care management for cattle and goat were also provided by the FET Group and CIRG to the villagers.

The findings through PRA techniques were triangulated with the villagers and the necessary feedback and corrections were taken up. The top most researchable problem we identified by discussions and using the ranking technique were triangulated and almost all the villagers were in total agreement on that.

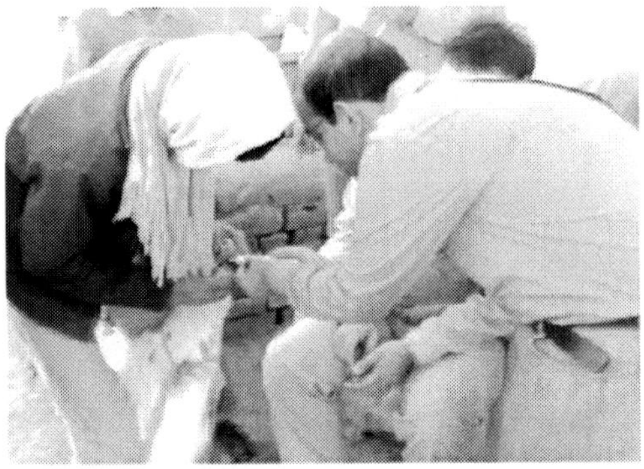

Plate 15. Goat oral medication during Health Camp

The broad issues we took up in the village seminar were the irrigation with saline water, soil testing, pest management, winter management, artificial insemination and vaccination in livestock and alternative livelihood employment opportunities from apiculture, mushroom cultivation, poultry and goat farming.

The farmers of Bar ka Nagla village perceived that saline water for drinking and irrigation was the main problem in the village. The FET group were able to explain the villagers that how to handle those problems. Therefore, there is a wide scope for diversifying the farming systems, which can improve the soil health and enhance the farmers' income and their living standards.

The seminar ended on a high note with the assurance that farmers will get advice for the cattle and goat related problems from the Health Division of CIRG, Makhdoom and Veterinary Hospital, Farah; soil and crop problems from the KVK, Mathura.

❑❑❑

Chapter 8

Institute Seminar

The institute seminar was held on 24th November, 2011 at CIRG, Makhdoom. Director Dr. D. Swarup, our FET Programme Co-ordinators Dr. Ashok Kumar and Dr. Vijay Kumar, and all the Scientists of CIRG, Makhdoom were present in the seminar. During the institute seminar, we discussed the village transect, resource and social maps, daily activity profile of both men and women, time line, seasonal calendar, consequence and venn diagrams, problem tree, solution tree and proposed research projects. The presentation by each FOCARs trainee, apart from different PRA tools also included information acquired about the problems of farmers.

Plate 16. FET Institute seminar at CIRG, Makhdoom

We presented the researchable issues to overcome the problem of pest and diseases in wheat, potato, bajra and mustard crop, animal health related problems and soil related problems. We triangulated the details that were collected from the village with the help of our

FET programme co-ordinators. After the seminar, an active discussion was held among Scientists and feedbacks were received. The Scientists gave valuable inputs to overcome most of the problems faced by the villagers, especially related to goat and cattle.

The dignitaries were happy with the FET group presentation and the PRA work. It was also suggested that further "farmer based - need oriented" research programmes should be initiated.

□□□

Chapter 9

Industrial Visits

Keeping in consideration, the socio-economic status of the villagers and to better understand the linkages of their occupation with the industry, we visited Panchhi Petha, and Virola International.

9.1 Visit to Panchhi Petha, Agra

Key Informants : Amar Singh, Product Manager

We visited the famous Petha making unit in Agra i.e. Panchhi Petha on 17th November, 2011. This is the petha making unit, where they used Ashgourd (Kachha petha) as raw material. They procure it from Etawah, Bareilly, and nearby districts of UP. Also, during lean season, they get the raw material from Mumbai and Chennai too.

The shelf life for their products is 4-5 days for wet petha and 15 days for dry petha. The manpower required for their manufacturing of 135 kg from 8 quintals of kaccha petha, 4-5 man per quintal. The whole petha preparation takes about 3-4 hours a day. The average cost benefit ratio (CBR) for petha making is 1:2. The villagers of the Bar ka Nagla can adapt the ashgourd cultivation as their livelihood income, which can be directly or indirectly linked to the petha making industry.

Manufacturing process involved in petha making is shown below.

Peeling of the ashgourds into small pieces

⇩

Wash with lime water, Boil for 1-2 hours to soften these

⇩

Dry them

⇩

Soak them in jaggery solution & cut into pieces

⇩

Dry them and packed in vacuum dry chamber with sugar

Plate 17. Visit to Panchhi Petha, Agra

9.2 Visit to Virola International, Agra

Key informants : Mr. R.D. Goyel, Production Manager

We visited Virola International, Agra on 17th November, 2011. This is the leather industry, mainly processed leather shoes are made for children, men and women. Around 1000 people were working in this industry out of which 100 belong to leather technology. It is 100 per cent export company and most of the leather shoes made by this company were exported to different European countries, like Italy, Germany & some African countries.

Plate 18. Visit to leather industry, Virola International, Agra

Value chain

- Processed animal skin (cow, sheep and goat) was the raw material and it was imported from Kolkata and Chennai.
- Imported leather is sent to SGS testing laboratory for checking important testing parameters, like tensile strength, abrasion resistance, etc. After getting satisfactory result, manufacturing process was started.
- Leather was cut by machines in different shapes.
- Stitching or Lasting process: In this process, cut leather was stitched with the upper part of shoes.
- Pasting: Sole was pated with the upper part of shoes.

Complete shoes were finished with different type of chemicals according to end use.

Chapter 10

Visit to Govt. Organizations and NGO

10.1 Visit to Krishi Vigyan Kendra, Mathura

To get acquainted with Cropping and animal husbandry scenario of Mathura District, we visited Krishi Vigyan Kendra, Mathura on 15 November, 2011 as a part of Line department Visits. Krishi Vigyan Kendra is working under administrative jurisdiction of U.P.Pt. Deen Dayal Upadhyaya Pashu Chikitsa Vigyan Vishwa Vidyalaya Evam Go-Anusandhan Sansthan, Mathura. It was established in 1986 with mandate of technology diffusion (Extension) and training to farmers & extension personnel of the district. The KVK has conducted Front Line Demonstration (FLD) and On Farm Testing (OFT) of technologies in Mustard, Potato, Wheat, Paddy crop and animal husbandry. Dr. R.K. Mishra, Programme Co-ordinator of the KVK, explained us about the mandates of KVK, cropping systems, land use pattern, technology adoption and agricultural problems, faced by the farmers of the district. We appraised him about the agricultural and animal husbandry problems identified in Bar Ka Nagla village. Problems, like Mustard aphid, Potato Frost injury, Salt affected soil, wheat termite, Bajra smut and goat diseases were discussed. Locality specific interventions against these problems were discussed by the programme co-ordinator which was suggested to farmers of the Bar ka Nagla village during village seminar.

Plate 19. Visit to KVK, Mathura

10.2 Visit to Central Soil and Water Conservation Research and Training Institute (CSWCRTI) Regional Centre, Agra UP

We visited CSWCRTI, Agra on 17 November, 2011. It is one among the eight sub-centres of CSWCRTI, Dehradun. The institute is working on reclamation of ravine lands of Yamuna river bank. We discussed with Dr. K.K Sharma, Incharge of the day about the measures to reduce the erosion of soil and conservation of water in this belt. He explained techniques that can be easily adopted by the farmers of this region to increase the infiltration and percolation of water into soil. He also suggested for reforestation of barren and degraded land with suitable trees species.

Plate 20. Visit to CSWCRTI, Regional Centre, Agra

10.3 Visit to Directorate of Rapeseed Mustard Research (DRMR), Bharatpur (Rajasthan)

Mustard is one of the major Rabi crop cultivated in Bar ka Nagla village. During transact walk and problem identification with matrix ranking, Aphid and salinity problems were given top priority by the farmers. In this context, we visited Directorate of Rapeseed Mustard on 18 November, 2011 with an objective to study the technologies developed by the centre for Mathura region.

Dr. Vinay Kumar, Scientist of the institute, briefed us about the organization and its mandates. The institute was first established as National Research Center on Rapeseed-Mustard (NRCRM) on October 20, 1993, which was re-designated as the Directorate of Rapeseed-mustard Research (DRMR) in February 2009 to carry out basic, strategic an applied research on Rapeseed and Mustard. The directorate has the mandate of planning, co-ordination and execution of research programme through wide network of 11 main, 12 sub and 22 verification centers across the country to augment production and productivity of Rapeseed-Mustard.

Plate 21. Visit to DRMR, Bharatpur (Rajasthan)

The DRMR developed suitable interventions for aphid and salinity management in our country. They are:

- Split application of Phosphorus in two doses in salinity affected soil.
- Cultivation of salt tolerant varieties, like CS-54, CS-52, Narendra Rai and BPR-1454. Spray of 2 per cent neem oil and 5 per cent NSKE for management of aphids.
- Spray of 0.02 per cent Oxydemeton Methyl and release of *Coccinella septum punctata* @ 5000 beetles /ha. For management of mustard aphids.

10.4 Visit to Central Institute for Research on Goats (CIRG) Laboratories and Facilities

We visited various divisions of CIRG viz. Nutrition, Feed Resource & Technology Division; Physiology & Reproduction Division; Genetics & Breeding Division; and Health Division on 19th November, 2011. The Nutrition Division deals with development of economic feeding system for goats and feed processing technologies for small holder and commercial goat farming system. Reproductive Division is working on goat Embryo transfer and *Invitro* fertilization technologies. Genetics division has significantly improved the productivity of Jamunapari and Barbari goats, besides supplying superior breeding bucks to different agencies for breed improvement programmes across the country. The Health Division are presently working on herbal medicine to control various diseases of goats, like Coccidiosis, Ectoparasites, etc. The Health Division also developed various molecular and serological diagnostic tools for detection of Caprine Brucellosis, Johne's disease and Mycoplasma infection in goats. We have also visited to the Extension Division of CIRG, which is dealing with studies of socioeconomic of goat production under different farming systems and transfer of viable production technologies to the field. We have also visited to CIRG farms, which consist of Barbari, Jamunapari, Jakhrana, Muzaffanagari sheep and experimental goat farms which consist of around 3000 goats.

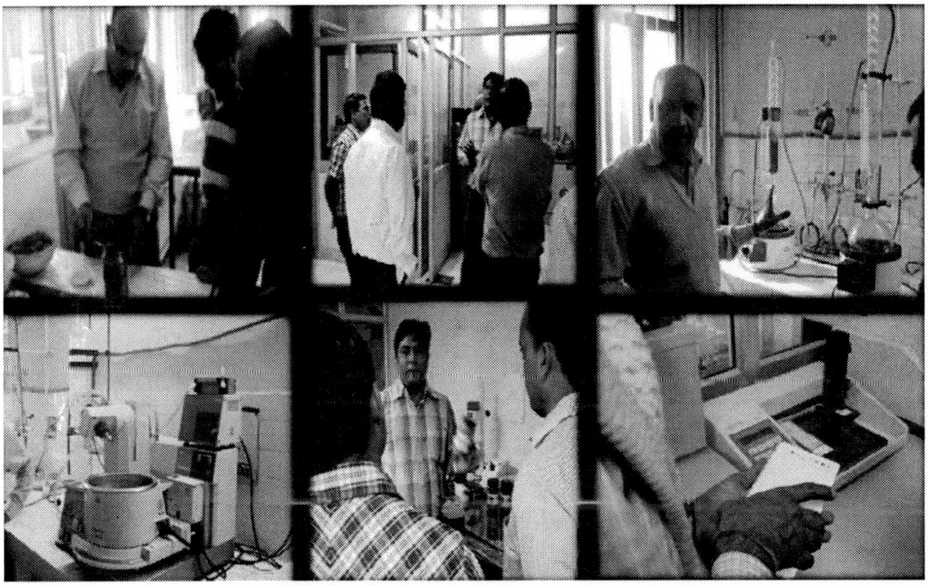

Plate 22. Visit to CIRG laboratories

10.5 Visit to NGO, Pt. Deendayal Dham, Nagla Chandrabhan, Farah, Mathura

We visited Pt. Deendayal Dham, located in Nagla Chandrabhan on 22 November, 2011. Shri Rameshpal Singh, Chairman of the NGO, explained us about the organization and its activities. The NGO was established as a memorial at Chandrabhan, which is the birthplace of late Pt. Deendayal with a mission to work for the welfare and development of society. Pt. Deendayal Dham Memorial NGO has profound influence on surrounding villages including Bar Ka Nagla village with its community development programmes. The NGO has conducted trials to identify vegetables and other crops suited for cultivation under briny conditions of surrounding villages. One of the cottage industries run by the dham is manufacturing Ayurvedic preparations, like Chyavanprash, Soaps and Cow urine based medicinal formulations. These cottage industries are also source of livelihood for Women from Bar Ka Nagla village. Apart from this, it is imparting vocational training to rural women of surrounding villages, like beekeeping, sewing, candle preparation and food processing. In addition, the memorial runs a charitable dispensary for treating ailments of poor rural people.

Plate 23. Visit to NGO, Pt. Deendayal Dham, Chandrabhan Nagla, Farah

❑❑❑

Chapter 11

Information and Communication Technology

In Bar ka Nagla village, the importance or awareness on Information and Communication Technology (ICT) related to agriculture is increasing but not to an adequate level, and still most of the farmers were not familiar with new technologies. ICT involves the conceptualization, design, development, evaluation and application of innovative ways to use information and communication in the rural domain, with a primary focus on agriculture and livestock. The leading information communications technologies, existing in the village were newspaper, radio, television, computer and mobile. Newspapers, like "Amar Ujala and Dainik Jagran" are available in the village are being subscribed by a few progressive farmers of the village. Some of them use their mobile to know the daily market prices, livestock and agricultural related problems and their solutions. Farmers are watching televisions for entrainment and also, agricultural related programme, like "Krishi Darshan". These programmes help them to acquire knowledge related to agriculture and animal husbandry. They use mobile for enquiring goat related problems, regarding pest and diseases with the CIRG, Makhdoom. They also use mobile for enquiring the price of seeds, fertilizers and also, ongoing market price of their product.

❑❑❑

Chapter 12

Exposure of FET Programme

Firstly, we have communicated to local newspaper about "Pashu Chikitsa Shivir" and "Kisan-Vaigyanik Gramin Sangosthi" held on 21st November, 2011 at Bar ka Nagla village. And we also shared our views and experiences of FET programme through ETV.

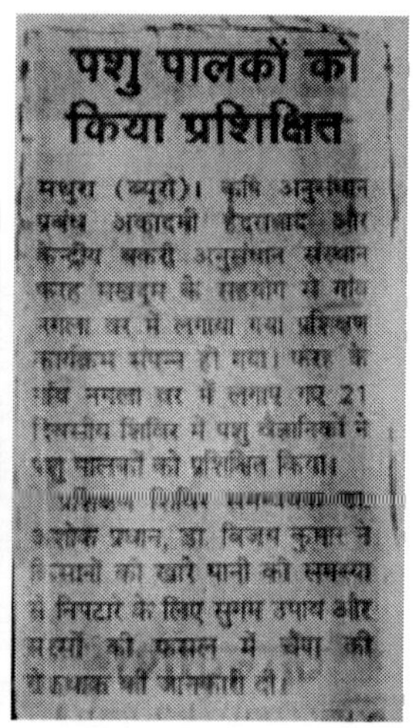

पशु पालकों को किया प्रशिक्षित

मथुरा (ब्यूरो)। कृषि अनुसंधान प्रबंध अकादमी हैदराबाद और केन्द्रीय बकरी अनुसंधान संस्थान फराह मखदूम के सहयोग से गांव नगला बर में लगाया गया प्रशिक्षण कार्यक्रम संपन्न हो गया। फरह के गांव नगला बर में लगाए गए 21 दिवसीय शिविर में पशु वैज्ञानिकों ने पशु पालकों को प्रशिक्षित किया।

प्रशिक्षण शिविर समन्वयक डा. अशोक प्रधान, डा. विजय कुमार ने किसानों को खारे पानी की समस्या से निपटारे के लिए सुगम उपाय और सरसों की फसल में चैंपा की रोकथाम की जानकारी दी।

Plate 24. Interaction with ETV – UP & UK News channel and published in local Newspaper (Amar Ujala)

❑❑❑

Chapter 13

Conclusions

Bar ka Nagla is a village, located at a distance of 4 km from our FET centre, CIRG, Makhdoom. Most of the population of this village were farmers and nearly 90 per cent villagers are involved in agriculture and livestock. This village comes under the potato-growing belt and thus, a majority of the crop produce was potato. Apart from potato, farmers were also taking wheat, bajra, mustard as major crops. In Rabi season, mustard, wheat, potato and berseem occupy a major area. In Khariff season, Dhaincha and Bajra are cultivated. No pulse crops were cultivated in the villages due to high salinity in soil and water, as pulses are highly sensitive to salinity. Most of the farmers were indulging in goat and cattle farming, but one farmer does sheep rearing and piggery.

Livestock rearing is also widespread in the village with few villagers were having cattle and goat. Landless villagers resort to goat rearing due the influence of CIRG, Makhdoom and some farmers also started doing backyard poultry. Goat's meat (chevon) and its milk (due to its medicinal traits) are also gaining wide importance in the global market. Goats have also been stated as the "future animal for rural prosperity". Being prolific eaters, goats can survive on almost on anything and graze usually on common property resources, which makes goat rearing a non-capital intensive occupation. Goat rearing is easy to start, requires low investment, provides high and consistent returns and a near liquid monetary status to the rearer. The net returns from commercial goat farming range between Rs. 970 to Rs. 2070 per goat per annum. With these benefits in mind and a little technical know-how from CIRG in its vicinity, the villagers of Bar ka Nagla can find an alternative and/or additional source of income from goat rearing.

As a major part of the cropped area is under potato and mustard crop in our study village, therefore, we have suggested that the mustard farmers in the village to go for potato chips to avoid wastage in leap season and provide extra income to the farmers, and also beekeeping as an additional activity. The potato growers of Bar Ka Nagla village should utilize the non-marketable potatoes through indigenous production of french fries and potato chips. Local non-government organizations, such as Deendayal Dham of Chandrabhan Nagla can provide the farmers with appropriate trainings for the same. Also, beekeeping can be practiced by anyone and requires low investment besides, providing high economic returns. It doesn't bring any pressure on agricultural land and produces honey, beeswax, pollen, propolis from the flowers which otherwise dry up in nature and go waste. Honeybees are not only important for their produce but also of greater importance to the farmers, as they play an important role in cross-pollinating various agricultural and horticultural crops and increase their yield per unit area and improve their quality. These products can be produced and further marketed in the nearby villages/ towns.

The situation in the field seems to be quite miserable due to saline water and saline soil, as we take a closer look at the village life. The village has saline water after the canal water stopped in the village. Due to this, several problems arose in the crops as well as in the livestock. The yield was drastically decreased, both in the crops and livestock. Therefore, we have suggested the farmers to used drip and micro-sprinkler irrigation system with filters to reduce salt content of irrigation water with minimum water utilization.

Inertia is inherent in any system and it is natural for farmers to continue doing what they have been practising for years. Farmers think more in terms of their immediate needs and situations. The long-term view is not the topmost priority for farmers. Hence, it is the scientists' obligation to create awareness among farmers of the long term effects of different cropping systems and how they should be more dynamic to keep pace with changes in the system. Creating awareness is also not an easy task and needs to be approached sensitively and sensibly with good planning. This FET and PRA exercise has made us more aware of the ground reality and has given us a better understanding of how to approach farmers and empower them to face the days ahead.

This Field Experience Training (FET) report has been prepared by multi-disciplinary team of eight members from different disciplines, namely Farm Machinery & Power, Plant Genetics, Plant Pathology, Agronomy, Horticulture, Animal Biotechnology, Agricultural Economics and Textile Chemistry. The entire scientist had lot of experience from this work. We can conclude, like 'For whom we are working, for whom we have to serve' through ICAR.